Advances in Neuroimaging Data Processing

Advances in Neuroimaging Data Processing

Editors

Alexander E. Hramov
Alexander N. Pisarchik

MDPI • Basel • Beijing • Wuhan • Barcelona • Belgrade • Manchester • Tokyo • Cluj • Tianjin

Editors
Alexander E. Hramov
Baltic Center for Artificial
Intelligence and
Neurotechnology, Immanuel
Kant Baltic Federal University
Kaliningrad
Russia

Alexander N. Pisarchik
Center for Biomedical
Technology, Universidad
Politécnica de Madrid,
Campus Montegancedo,
Pozuelo de Alarcón
Madrid
Spain

Editorial Office
MDPI
St. Alban-Anlage 66
4052 Basel, Switzerland

This is a reprint of articles from the Special Issue published online in the open access journal *Applied Sciences* (ISSN 2076-3417) (available at: https://www.mdpi.com/journal/applsci/special_issues/Neuroimaging_Data_Processing).

For citation purposes, cite each article independently as indicated on the article page online and as indicated below:

LastName, A.A.; LastName, B.B.; LastName, C.C. Article Title. *Journal Name* **Year**, *Volume Number*, Page Range.

ISBN 978-3-0365-6998-7 (Hbk)
ISBN 978-3-0365-6999-4 (PDF)

© 2023 by the authors. Articles in this book are Open Access and distributed under the Creative Commons Attribution (CC BY) license, which allows users to download, copy and build upon published articles, as long as the author and publisher are properly credited, which ensures maximum dissemination and a wider impact of our publications.

The book as a whole is distributed by MDPI under the terms and conditions of the Creative Commons license CC BY-NC-ND.

Contents

About the Editors vii

Preface to "Advances in Neuroimaging Data Processing" ix

Alexander. E. Hramov and Alexander. N. Pisarchik
Special Issue "Advances in Neuroimaging Data Processing"
Reprinted from: *Appl. Sci.* **2023**, *13*, 2060, doi:10.3390/app13042060 1

Axel Faes, Iris Vantieghem and Marc M. Van Hulle
Neural Networks for Directed Connectivity Estimation in Source-Reconstructed EEG Data
Reprinted from: *Appl. Sci.* **2022**, *12*, 2889, doi:10.3390/app13042060 5

Christina Maher, Arkiev D'Souza, Michael Barnett, Omid Kavehei, Chenyu Wang and Armin Nikpour
Structure-Function Coupling Reveals Seizure Onset Connectivity Patterns
Reprinted from: *Appl. Sci.* **2022**, *12*, 10487, doi:10.3390/app122010487 25

Xiaobi Chen, Guanghua Xu, Sicong Zhang, Xun Zhang and Zhicheng Teng
Building Networks with a New Cross-Bubble Transition Entropy for Quantitative Assessment of Mental Arithmetic Electroencephalogram
Reprinted from: *Appl. Sci.* **2022**, *12*, 11165, doi:10.3390/app122111165 39

Alexey N. Pavlov, Alexander I. Dubrovskii, Olga N. Pavlova and Oxana V. Semyachkina-Glushkovskaya
Effects of Sleep Deprivation on the Brain Electrical Activity in Mice
Reprinted from: *Appl. Sci.* **2021**, *11*, 1182, doi:10.3390/app11031182 55

Parth Chholak, Semen A. Kurkin, Alexander E. Hramov and Alexander N. Pisarchik
Event-Related Coherence in Visual Cortex and Brain Noise: An MEG Study
Reprinted from: *Appl. Sci.* **2021**, *11*, 375, doi:10.3390/app11010375 65

Iván De La Pava Panche, Andrés Álvarez-Meza, Paula Marcela Herrera Gómez, David Cárdenas-Peña, Jorge Iván Ríos Patiño and Álvaro Orozco-Gutiérrez
Kernel-Based Phase Transfer Entropy with Enhanced Feature Relevance Analysis for Brain Computer Interfaces
Reprinted from: *Appl. Sci.* **2021**, *11*, 6689, doi:10.3390/app11156689 77

Alexander K. Kuc, Semen A. Kurkin, Vladimir A. Maksimenko, Alexander E. Hramov and Alexander N. Pisarchik
Monitoring Brain State and Behavioral Performance during Repetitive Visual Stimulation
Reprinted from: *Appl. Sci.* **2021**, *11*, 11544, doi:10.3390/app112311544 103

Arne Van Den Kerchove, Arno Libert, Benjamin Wittevrongel and Marc M.Van Hulle
Classification of Event-Related Potentials with RegularizedSpatiotemporal LCMV Beamforming
Reprinted from: *Appl. Sci.* **2022**, *12*, 2918, doi:10.3390/app12062918 117

Ba-Viet Ngo and Thanh-Hai Nguyen
A Semi-Automatic Wheelchair with Navigation Based on Virtual-Real 2D Grid Maps and EEG Signals
Reprinted from: *Appl. Sci.* **2022**, *12*, 8880, doi:10.3390/app12178880 135

Miguel Ángel Sánchez-Cifo, Francisco Montero and María Teresa López
MuseStudio: Brain Activity Data Management Library for Low-Cost EEG Devices
Reprinted from: *Appl. Sci.* **2021**, *11*, 7644, doi:10.3390/app11167644 . . . . **155**

About the Editors

Alexander E. Hramov

Alexander E. Hramov is a Professor and the Head of the Baltic Center for Neurotechnology and Artificial Intelligence at Immanuel Kant Baltic Federal University in Kaliningrad, Russia. His research focuses on complex systems theory, artificial intelligence, and graph theory as applied to biomedical and cognitive studies. Prof. Hramov has authored and edited 7 monographs, 27 patents, and approximately 250 peer-reviewed scientific papers. He serves as Editor for several journals, including *Chaos, Solitons and Fractals, Applied Sciences, Russian Journal of Nonlinear Dynamics, and Information and Control System,* and is an Advisory Editorial Board Member for *Heliyon*. He is also a Board Member of the International Physics and Control Society (IPACS).

Alexander Pisarchik

Alexander Pisarchik is a Distinguished Researcher and Isaac-Peral Chair in Computational Systems Biology at the Center for Biomedical Technology of Universidad Politécnica de Madrid. His research interests include nonlinear and stochastic dynamics of biomedical and physical systems, EEG, MEG, cognitive neuroscience, neuronal models, and brain–computer interfaces. He has authored and served as editor for 5 monographs, 18 book chapters, 13 patents, and about 300 papers in peer-reviewed scientific journals. Dr. Pisarchik is Associate Editor of *Applied Sciences, Frontiers in Network Physiology,* and the *International Journal of Discontinuity, Nonlinearity and Complexity*, Academic Editor of *PLoS One*, Editor-in-Chief of *Advances in Biology, Biotechnology and Genetics*, Section Editor of *Open Life Sciences*, and an editorial member of several journals. He is a Board Member of the International Physics and Control Society (IPACS) and a member of Consorcio de Investigación Biomédica en Red (CIBER).

Preface to "Advances in Neuroimaging Data Processing"

The Special Issue "Advances in Neuroimaging Data Processing" collects original research papers highlighting the recent advances in neuroimaging-based data processing using novel experimental and computational approaches. Various topics are covered herein, including the development of the novel methods of EEG and MEG brain activity processing, the restoration and analysis of functional brain networks, the creation of brain–computer interfaces (BCIs), and the external device neurocontrol.

Alexander E. Hramov and Alexander N. Pisarchik
Editors

Editorial

Special Issue "Advances in Neuroimaging Data Processing"

Alexander. E. Hramov [1,*] and Alexander. N. Pisarchik [2,*]

1. Baltic Center for Neurotechnology and Artificial Intelligence, Immanuel Kant Baltic Federal University, Kaliningrad 236041, Russia
2. Center for Biomedical Technology, Technical University of Madrid, Campus Montegancedo, 28223 Madrid, Spain
* Correspondence: aekhramov@kantiana.ru (A.E.H.); alexander.pisarchik@ctb.upm.es (A.N.P.)

1. Introduction

The development of in vivo neuroimaging technology has led to an incredible amount of digital information concerning the brain. Neuroimaging techniques are being increasingly used to study human cognitive processes [1] and create brain–machine interfaces [2], as well as to identify and diagnose certain brain disorders [3]. Currently, neuroscientists and physicians actively use various methods of brain scanning, including electroencephalography (EEG), magnetoencephalography (MEG), functional near-infrared spectroscopy (fNIRS), electrocorticography (ECoG), functional magnetic resonance imaging (fMRI), positron emission tomography (PET), and diffusion tensor imaging (DTI). Recent advances in signal processing and machine learning applied to neuroimaging data using various signal-processing methods have led to impressive progress towards solving several practical problems in medicine, healthcare, neuroscience, biomedical engineering, brain–machine interfaces, cognitive science, etc.

2. Advanced Methods of Neuroimaging-Based Data Processing

In light of the foregoing discussion, this Special Issue collects original papers on theoretical and experimental results highlighting the recent advances in neuroimaging-based data processing using theories, algorithms, architectures, and applications. Various topics are covered herein, mainly those related to the restoration of functional brain networks, the processing of EEG and MEG brain activity, brain state monitoring including brain–computer interfaces (BCIs) and external device control, and the development of open-source software tools.

One of the most important tasks of neuroimaging is the restoration of functional brain networks [4,5]. Therefore, three out of ten articles in this Special Issue are devoted to the analysis and development of methods for restoring functional brain networks [6–8]. In the first paper, Faes, Vantieghem, and Van Hulle [6] propose a new approach to source reconstruction from EEG data. Their method is based on directed connectivity estimation using deep learning. The authors apply several types of artificial neural networks to estimate directed connectivity and assess its accuracy with respect to several ground truths. They show that an LSTM neural network with non-uniform embedding yields the most promising results due to its relative robustness to differing dipole locations. In the second paper, Maher et al. [7] discuss the application of multimodal neuroimaging data to revealing connectivity patterns in the onset of seizures. The authors obtain structural connectomes from diffusion MRI (dMRI) and functional connectomes from EEG to assess whether high structure–function coupling corresponds to the seizure onset region. They argue that dMRI combined with EEG can improve the identification of the seizure onset region. Their study is a good example of dMRI's potential in clinical practice. Finally, in the third paper, Chen et al. [8] present a new method for constructing complex networks to assess cognitive load. Their approach is based on cross-permutation entropy.

Traditional research tasks in the field of neuroimaging include time series analysis of brain activity recordings. In this regard, Pavlov et al. [9] discuss the possibility of detecting changes in brain electrical activity associated with sleep deficit using extended detrended fluctuation analysis (EDFA). By applying this approach to EEGs in mice, they identify signs of changes that could be caused by short-term sleep deprivation. In another paper, Chholak et al. [10] analyze the neurophysiological data of MEG experiments based on the visual perception of flickering ambiguous stimuli. The results support their hypothesis of a correlation between event-related coherence in the visual cortex and neuronal noise and suggest that greater brain involvement in visual stimuli is accompanied by stronger brain noise.

Several papers in this Special Issue involve the active development of neurotechnology. Specifically, in their paper, De La Pava Panche et al. [11] present a new method for estimating phase transfer entropy (TE) between distinct pairs of instantaneous phase time series to enable real-time estimation, which is important for the development of TE application strategies for BCIs. In another paper, Kuc et al. [12] propose a monitoring system that facilitates the evaluation of behavioral performance (decision time and errors) during a prolonged visual classification task. The results of this work enable the determination of whether changes in pre-stimulus neural activity, as measured by EEG power, predict behavioral characteristics. In the following article, Van Den Kerchove et al. [13] consider the problem of the usability of EEG-based visual BCIs using event-related potentials (ERPs). They introduce two regularized estimators for beamformer weights that are well conditioned despite using limited training data and improve ERP classification accuracy to reduce calibration time and required EEG data before BCI operation. Along with the rapid development and exploitation of methods for the real-time processing of brain activity, these estimators' roles in solving the problems of the rehabilitation of patients suffering from brain injuries or helping people with disabilities are becoming essential, as also highlighted in the paper by Ngo and Nguyen [14], who propose an EEG-based wheelchair control system using a grid map designed to enable people with disabilities to reach any given destination.

Finally, Sánchez-Cifo, Montero, and López [15] describe a developed open-source tool called MuseStudio that allows one to import and export data from brain-sensing headband EEG devices and view and analyze brain data in real time.

3. Future of Neuroimaging Data Processing: Is the Era of Machine Learning Coming?

In recent years, significant progress has been achieved regarding the methods of analysis and processing of neuroimaging data, primarily through the use of machine learning. Recently, machine learning has gained popularity in neuroscience due to its ability to recognize hidden patterns and nonlinear relationships in large volumes of nonstationary and ambiguous neuroimaging data. Soon, biologists and mathematicians can anticipate the greater use of machine learning approaches to gain new insights into brain behavior and neurotechnology applications, including BCIs. Machine learning is of particular interest for the medical diagnosis of neurological diseases, where it is a powerful tool for the early detection of biomarkers of various neurological disorders [16]. At the same time, appropriate methods and approaches regarding explainable artificial intelligence (XAI) should be ready for the integration and use for neuroimaging-based data processing in modern digital medicine [17,18].

Funding: This research was funded by Program 'Priority-2030' of Immanuel Kant Baltic Federal University of Ministry of Education and Science of Russian Federation.

Acknowledgments: We thank all authors and peer reviewers for their valuable contributions to this Special Issue. We are also grateful to the MDPI management and staff for their tireless editorial support for this project, which led to its successful completion, and for the launch of an open-access international peer-reviewed journal, *Applied Sciences*, covering all aspects of applied natural science.

Conflicts of Interest: The authors declare no conflict of interest.

References

1. Morita, T.; Asada, M.; Naito, E. Contribution of Neuroimaging Studies to Understanding Development of Human Cognitive Brain Functions. *Front. Hum. Neurosci.* **2016**, *10*, 464. [CrossRef] [PubMed]
2. Hramov, A.E.; Maksimenko, V.A.; Pisarchik, A.N. Physical principles of brain–computer interfaces and their applications for rehabilitation, robotics and control of human brain states. *Phys. Rep.* **2021**, *918*, 1–133. [CrossRef]
3. Abi-Dargham, A.; Horga, G. The search for imaging biomarkers in psychiatric disorders. *Nat. Med.* **2016**, *22*, 1248–1255. [CrossRef] [PubMed]
4. Lynn, C.W.; Bassett, D.S. The physics of brain network structure, function and control. *Nat. Rev. Phys.* **2019**, *1*, 318–332. [CrossRef]
5. Hramov, A.E.; Frolov, N.S.; Maksimenko, V.A.; Kurkin, S.A.; Kazantsev, V.B.; Pisarchik, A.N. Functional networks of the brain: From connectivity restoration to dynamic integration. *Physics-Uspekhi* **2021**, *64*, 584–616. [CrossRef]
6. Faes, A.; Vantieghem, I.; Van Hulle, M.M. Neural Networks for Directed Connectivity Estimation in Source-Reconstructed EEG Data. *Appl. Sci.* **2022**, *12*, 2889. [CrossRef]
7. Maher, C.; D'Souza, A.; Barnett, M.; Kavehei, O.; Wang, C.; Nikpour, A. Structure-Function Coupling Reveals Seizure Onset Connectivity Patterns. *Appl. Sci.* **2022**, *12*, 10487. [CrossRef]
8. Chen, X.; Xu, G.; Zhang, S.; Zhang, X.; Teng, Z. Building Networks with a New Cross-Bubble Transition Entropy for Quantitative Assessment of Mental Arithmetic Electroencephalogram. *Appl. Sci.* **2022**, *12*, 11165. [CrossRef]
9. Pavlov, A.; Dubrovskii, A.; Pavlova, O.; Semyachkina-Glushkovskaya, O. Effects of Sleep Deprivation on the Brain Electrical Activity in Mice. *Appl. Sci.* **2021**, *11*, 1182. [CrossRef]
10. Chholak, P.; Kurkin, S.; Hramov, A.; Pisarchik, A. Event-Related Coherence in Visual Cortex and Brain Noise: An MEG Study. *Appl. Sci.* **2021**, *11*, 375. [CrossRef]
11. De La Pava Panche, I.; Álvarez-Meza, A.; Herrera Gómez, P.; Cárdenas-Peña, D.; Ríos Patiño, J.; Orozco-Gutiérrez, Á. Kernel-Based Phase Transfer Entropy with Enhanced Feature Relevance Analysis for Brain Computer Interfaces. *Appl. Sci.* **2021**, *11*, 6689. [CrossRef]
12. Kuc, A.; Kurkin, S.; Maksimenko, V.; Pisarchik, A.; Hramov, A. Monitoring Brain State and Behavioral Performance during Repetitive Visual Stimulation. *Appl. Sci.* **2021**, *11*, 11544. [CrossRef]
13. Van Den Kerchove, A.; Libert, A.; Wittevrongel, B.; Van Hulle, M. Classification of Event-Related Potentials with Regularized Spatiotemporal LCMV Beamforming. *Appl. Sci.* **2022**, *12*, 2918. [CrossRef]
14. Ngo, B.; Nguyen, T. A Semi-Automatic Wheelchair with Navigation Based on Virtual-Real 2D Grid Maps and EEG Signals. *Appl. Sci.* **2022**, *12*, 8880. [CrossRef]
15. Sánchez-Cifo, M.; Montero, F.; López, M. MuseStudio: Brain Activity Data Management Library for Low-Cost EEG Devices. *Appl. Sci.* **2021**, *11*, 7644. [CrossRef]
16. Rajpurkar, P.; Chen, E.; Banerjee, O.; Topol, E.J. AI in health and medicine. *Nat. Med.* **2022**, *28*, 31–38. [CrossRef]
17. Tjoa, E.; Guan, C. A survey on explainable artificial intelligence (xai): Toward medical xai. *IEEE Trans. Neural Netw. Learn. Syst.* **2020**, *32*, 4793–4813. [CrossRef] [PubMed]
18. Karpov, O.E.; Andrikov, D.A.; Maksimenko, V.A.; Hramov, A.E. Explainable Artificial Intelligence for Medicine. *Dr. Inf. Technol.* **2022**, *2*, 4–11. [CrossRef]

Disclaimer/Publisher's Note: The statements, opinions and data contained in all publications are solely those of the individual author(s) and contributor(s) and not of MDPI and/or the editor(s). MDPI and/or the editor(s) disclaim responsibility for any injury to people or property resulting from any ideas, methods, instructions or products referred to in the content.

Article

Neural Networks for Directed Connectivity Estimation in Source-Reconstructed EEG Data

Axel Faes *,†, Iris Vantieghem † and Marc M. Van Hulle

Laboratory for Neuro- and Psychophysiology, Department of Neurosciences, Medical School, Katholieke Universiteit Leuven, 3000 Leuven, Belgium; iris.vantieghem@student.kuleuven.be (I.V.); marc.vanhulle@kuleuven.be (M.M.V.H.)
* Correspondence: axel.faes@kuleuven.be
† These authors contributed equally to this work.

Abstract: Directed connectivity between brain sources identified from scalp electroencephalography (EEG) can shed light on the brain's information flows and provide a biomarker of neurological disorders. However, as volume conductance results in scalp activity being a mix of activities originating from multiple sources, the correct interpretation of their connectivity is a formidable challenge despite source localization being applied with some success. Traditional connectivity approaches rely on statistical assumptions that usually do not hold for EEG, calling for a model-free approach. We investigated several types of Artificial Neural Networks in estimating Directed Connectivity between Reconstructed EEG Sources and assessed their accuracy with respect to several ground truths. We show that a Long Short-Term Memory neural network with Non-Uniform Embedding yields the most promising results due to its relative robustness to differing dipole locations. We conclude that certain network architectures can compete with the already established methods for brain connectivity analysis.

Keywords: brain connectivity; artificial neural networks; source reconstruction; granger causality; time series

1. Introduction

A challenging problem in neuroimaging is to estimate directed connectivity between brain regions reconstructed from scalp EEG recordings but important to unveil their joint dynamics. Due to volume conduction, a given EEG electrode can pick up signals from several sources simultaneously, distorted along the way due to the presence of tissues with different electrical properties. Resolving these sources is called the "inverse problem", and it consists of estimating the source parameters given the scalp EEG recordings. The number of sources is higher than the number of electrodes, rendering an ill-posed problem. Valid brain connectivity estimation critically depends on the correct localization and time series reconstruction in this stage. Several localization methods have been proposed, often yielding differing outcomes. In a comprehensive set of simulations, [1] studied the influence of several inverse solutions, the depth of the sources, their reciprocal distance, and the Signal-to-Noise Ratio (SNR) of the recordings. They found that all these factors had a significant impact on the resulting connectivity pattern and that the number of spurious connectivity estimations depends heavily on the combinations of these factors.

In addition to the said factors, the choice of the connectivity estimator also has a significant impact. Our interest lies in directed connectivity estimation, of which partial directed coherence [2], dynamic causal modeling [3], structural equation modeling [4] and (conditional) Granger causality (GC) [5] are well-known methods. However, they rely on statistical assumptions that usually do not hold for EEG data, such as linearity [6], stationarity and prior assumptions on connectivity being expressible as a relation between

time series. However, even though some of these assumptions are violated, these methods still are best practice cases of directed connectivity estimation. In what follows, we focused on variations in traditional Granger Causality, given that it does not rely on an a priori assumed connectivity pattern. Granger Causality is a statistical hypothesis used to determine temporal causal effects between two time series. If the past of a second time series (Z) together with the past of a first time series (Y) (i.e., the "full" model) results in an improved prediction of the future value of the first time series, then the past of the first time series alone (the "reduced" model), it is said that time series Z "Granger-causes" Y.

Two main problems with this bivariate model can be discerned. Firstly, bivariate GC does not account for other time series that may be causing both Y and Z, resulting in spurious connectivity patterns. Secondly, even when bivariate GC is extended towards multiple time series by conditioning on these other variables, it is still possible that the found influence is actually caused by a linear mixture of non-interacting sources. This is because the signal measured from one electrode usually contains contributions of several sources [7]. Important to note is the proposal of Time-Reversed Granger Causality (TRGC) by [8], further validated by [7], to reduce the impact of additive correlational noise due to source mixing. The idea is that when connectivity is based on temporal delay, directed connectivity should be reversed when the temporal order is reversed. Concretely, it is checked whether the obtained GC scores for non-reversed and reversed data have opposing directions and are both significant [1]. This is clearly different from a classical way to determine significance (i.e., a likelihood ratio test). Hence, the main difference between TRGC and traditional GC is the proposed significance procedure. Still, even with TRGC, errors in connectivity estimation are here to stay. The question remains whether a totally different approach could cope with the above-mentioned problems and could perform better, or at least equally well, in comparison with the standard approaches. Artificial Neural Networks (ANNs) were considered as particularly interesting candidates given their flexible way of approximating highly non-linear relationships between variables [9] and the fact that no a priori assumptions need to be made about signal stationarity nor the connectivity pattern (for a clear overview, see [10]). Temporal convolutional networks (TCNs), as well as recurrent neural networks (RNNs), are usually well-suited architectures for time series [11–15]. While RNNs are often seen as the gold standard for sequence modeling, TCNs have also proven their suitability, for instance, in financial forecasting [14], electric power forecasting [16] and language modeling [17]. However, it remains unclear whether ANNs can signal the presence or absence of connections and their strength. Although some authors already used ANNs to derive directed brain connectivity with multilayer perceptrons and recurrent networks [15,18], these approaches did not include source-reconstructed EEG data. As stated before, unlike EEG source reconstruction, analyses based on EEG electrode levels do not allow for trustworthy inferences about interacting regions [19]. Hence, the suitability of ANNs in deriving directed connectivity between reconstructed EEG sources remains unknown.

Our motivation to assess ANNs for directed connectivity estimation between reconstructed EEG sources was two-fold. First, although many connectivity estimators exist, it is not yet known which current ANNs architectures can cope better with source-reconstructed EEG activity and under various circumstances. The authors of [1] were the first to conduct a comprehensive simulation study on the influence of dipole location, noise level, inverse solution and connectivity estimation, as well as the interactions between these factors. It was shown that different circumstances call for different analysis pipelines and that under advanced noise levels and for particular dipole configurations, even well-established methods such as TRGC can return aberrant connectivity estimates. Second, ANNs boast several appealing modeling properties that are potentially relevant to EEG modelers, such as the ability to deal with non-stationarity, non-linearity and, depending on the ANN architecture, to dispense with the prior specification of model order.

In order to assess the ability of ANNs to correctly signal the presence or absence of directed connectivity as well as connectivity strength, we compared several ANN models, including Conv2D, a novel ANN model we propose, with TRGC. We compared their performance for different dipole locations (i.e., Far–Deep/Far–Superficial) as this can inform us whether there is a future for ANN models in brain connectivity estimation. In addition, we evaluated the ANN models relevant for directed connectivity estimation. We investigated these issues by means of a simulation study, thereby making use of a slightly adapted version of the simulation framework developed by [1] in which we manipulated the location of the dipoles and their connectivity while keeping noise level and the choice of the inverse solution constant.

2. Materials and Methods

2.1. Simulation Procedure

The simulation framework developed by [1] was used to generate simulated EEG data originating from three dipoles. This data generating process, as well as the forward and inverse problems, were implemented in MATLAB (2020). Figure 1 shows the data generation procedure. The standard length of each generated series was 1500 time steps for Ground Truth 1 and 2.

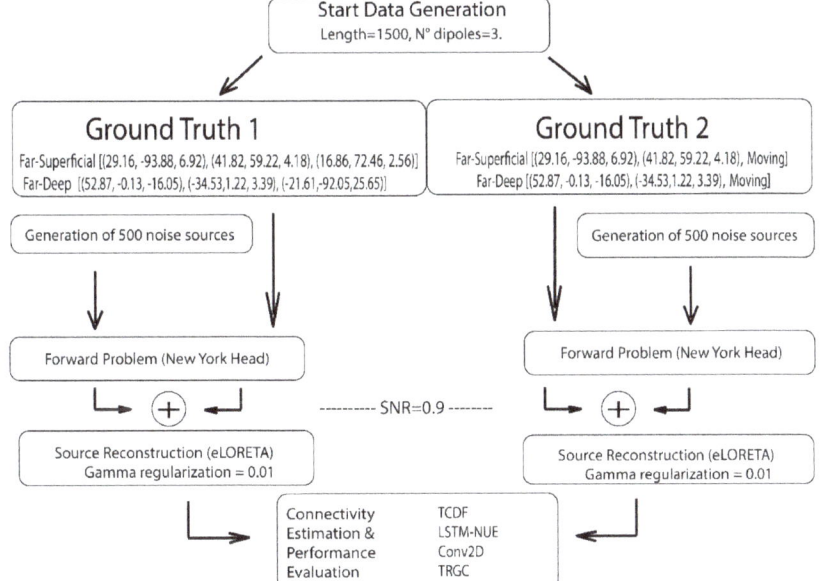

Figure 1. Simulation Procedure followed by Connectivity Estimation. TCDF = Depthwise Separable 1D Temporal Causal Discovery Framework, LSTM-NUE = Long Short-Term Memory with Non-Uniform Embedding, Conv2D = 2D Convolutional Network, TRGC = Time-Reversed Granger Causality. Coordinates in the Ground Truths denote MNI-coordinates.

In Ground Truth 1, three fixed dipoles were used with the directionality of the connections as well as their strength being imposed (Figure 2), a strategy used before [20–22]:

$$X_1(t) = 0.5X_1(t-1) - 0.7X_1(t-2) + c12(t)X_2 + \in_1(t)$$
$$X_2(t) = 0.7X_2(t-1) - 0.5X_2(t-2) + 0.2X_1 + c23(t)X_3(t-1) + \in_2(t) \quad (1)$$
$$X_3(t) = 0.8X_3(t-1) + \in_3(t)$$

with X_1, X_2 and X_3, three electrical sources contributing to the simulated scalp-EEG signals and with:

$$c12(t) = 0.5\tfrac{t}{L} \; if \; t \leq \tfrac{L}{2} \;, \; c12(t) = 0.5\tfrac{L-t}{\tfrac{L}{2}} \; if \; t > \tfrac{L}{2}$$
$$c23(t) = 0.4 \; if \; t < 0.7L, \; c23(t) = 0 \; if \; t \geq 0.7L \quad (2)$$

L = length of the generated time series (L = 1500), t = the current time step and ε = uncorrelated white noise, varying with time. We further assume an EEG cap with 108 electrodes.

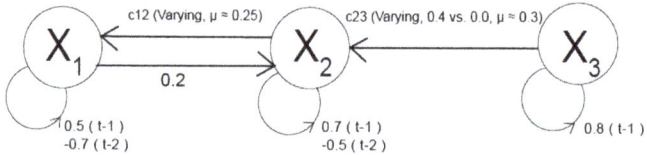

Figure 2. Ground Truth 1 with three fixed dipoles.

For Ground Truth 2, we considered two fixed, one moving dipole and only one true connection (Figure 3) and focused on the presence or absence of this connectivity as well as its directionality:

$$\begin{bmatrix} X_s(t) \\ X_r(t) \\ X_n(t) \end{bmatrix} = \sum_{p=1}^{P} \begin{bmatrix} a_{11}(p) & 0 & 0 \\ a_{21}(p) & a_{22}(p) & 0 \\ 0 & 0 & a_{33}(p) \end{bmatrix} \begin{bmatrix} X_s(t-p) \\ X_r(t-p) \\ X_n(t-p) \end{bmatrix} + \begin{bmatrix} \epsilon_1(t) \\ \epsilon_2(t) \\ \epsilon_3(t) \end{bmatrix} \quad (3)$$

with X_s, the moving dipole, as a sender, and two fixed dipoles, with X_r the receiver and X_n the fixed non-interactive dipole, and a_{ij} (p), i, j ∈ {1, 2, 3} and p ∈ {1, ..., P} the coefficients with a_{21} the coupling strength between sender and receiver. All a_{ij} are randomly picked from the interval [0.3, 1]. Finally, ϵ is uncorrelated, biological, white noise.

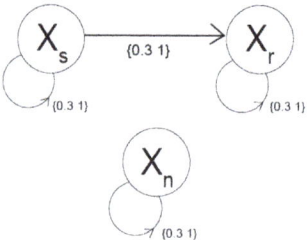

Figure 3. Ground Truth 2 with two fixed, one moving dipole.

The moving dipole (the sender) changes location (far, deep, close, superficial) at every iteration, with a total of 1004 iterations. The maximum time lag t is two. The reason for this ground truth is that the sender can be located at really challenging locations (too close to one of the other dipoles or very deep in the brain).

Two conditions were created for both ground truths: one condition consisted of three superficial dipoles far away from each other, while the other consisted of three dipoles located "deep" in the brain, but each dipole was still positioned far away from the other dipoles. The corresponding MNI-coordinates of the two fixed dipoles that Ground Truth 1 and 2 have in common are depicted in Figure 4. The full set of coordinates of Ground Truth 1 (including the coordinates of the third fixed dipole) is denoted in Figure 1.

In Ground Truth 2, the first two coordinates are the same as in Ground Truth 1, for each dipole condition, while the third dipole moves throughout the brain as described above. The Far–Superficial versus Far–Deep configurations indicate (relative) distances: "deep" denotes a distance from the origin (located at the anterior commissure) <6 cm and "superficial" >6.5 cm. The distance between dipoles is evaluated as "far" if the relative distance to the other dipoles exceeds 8 cm.

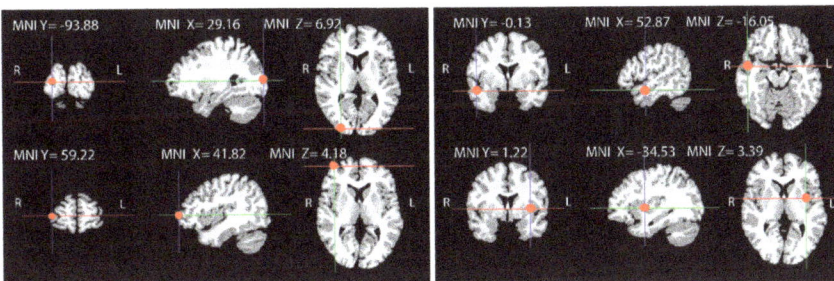

Figure 4. Locations (MNI-coordinates) of the dipoles the Ground Truths have in common. Left: Far–Superficial fixed dipoles. Right: Far–Deep fixed dipoles.

As an additional check for robustness of source localization, noise sources were added as a background activity. These were modeled using pink noise, also called 1/f noise, and created by scaling the amplitude spectrum of random white Gaussian noise with the factor 1/f using the Fourier transform and its inverse.

After generating these noise sources, the forward problem is construed:

$$Y = LX + e, \qquad (4)$$

where Y denotes the scalp-recorded potentials, X represents the electrical sources in the brain (the dipoles), "e" is measurement noise (electrode noise) and L is the head volume conductor model (also called the leadfield matrix). The leadfield matrix determines how the activity flows from dipoles to electrodes. In this work, the New York Head model [23] was used.

The pink noise and the source activity are then projected onto the scalp, after which they are summed:

$$Y^{brain}(t) = \gamma \times \frac{Y^{active}(t)}{||Y^{active}(t)||_{FRO}} + (1-\gamma) \times \frac{Y^{noise}}{||Y^{noise}(t)||_{FRO}} \qquad (5)$$

Y^{active} and Y^{noise} refer to the scalp-projected source signals and pink noise activity, respectively; both are scaled by dividing them by their Frobenius norm ($||Y^{active}(t)||_{FRO}$, $||Y^{active}(t)||_{FRO}$). The Signal-to-Noise Ratio (SNR) is computed for all dipoles simultaneously and set to 0.9 ($\gamma = 0.9$).

Next, white noise (spatially and temporally uncorrelated activity) is added to Y^{brain} to simulate electrode noise, resulting in Equation (6) where $Y^{measurement}$ represents the simulated EEG signal. Again $\gamma = 0.9$ is imposed as Signal-to-Noise Ratio:

$$Y^{measurement}(t) = 0.9 \times \frac{Y^{brain}(t)}{||Y^{brain}(t)||_{FRO}} + 0.1 \times \frac{Y^{meas\_noise}}{||Y^{meas\_noise}(t)||_{FRO}} \qquad (6)$$

Afterwards, the simulated scalp-EEG data are source-reconstructed using exact low-resolution brain electromagnetic tomography (eLORETA) [24]. There have also been improvements to eLORETA, such as Sparse eLORETA, which uses a masking approach to improve the source localization density [25]. The eLORETA method is a discrete, three-dimensional (3D), linear, weighted minimum norm inverse solution [24]. In the absence of noise, an exact zero-error localization accuracy can be obtained with eLORETA, but this does not hold for noisy data, as was shown in a study comparing both scenarios [26]. The MATLAB implementation of the eLORETA algorithm (mkfilt_eloreta2.m) from which spatial filters are obtained was developed by G. Nolte and is available in the MEG/EEG Toolbox of Hamburg (METH; https://www.uke.de/english/departments-institutes/institutes/neurophysiology-and-pathophysiology/research/research-groups/index.html, accessed on 22 December 2021). As the input, it takes the leadfield tensor (i.e., the head model file N*M*P containing N channels, M voxels, and P dipole directions) as well as a regularization parameter gamma (set to 0.01); as the output, an N*M*P tensor A of spatial filters is returned.

2.2. Connectivity Models

The ANNs evaluated in this study were selected based on their suitability for time series analysis (Table 1). While TCDF outputs attention scores, for which higher scores are used to represent stronger connectivities, LSTM-NUE makes use of Granger Causality scores equaling NNGC = errreduced−errfull, which are then binarized [18]. Conv2D uses the R^2-score between the real and predicted values of the current target (i.e., the time steps to be predicted). While using TRGC as implemented by [1], only binary GC scores are outputted. The configuration (i.e., the used parameters) of each ANN was determined using a data-driven approach, such that for each ANN, the parameters returning the best results were chosen. This parameters pre-testing was performed with different simulated data sets (i.e., differing from the data sets that were used to report the final results).

Table 1. Set-up of the compared ANNs and TRGC.

Model	Architecture	Self-Causation	Connectivity Measure
TCDF	Depthwise Separable 1D Convolutional Network	yes	Attention score
LSTM-NUE	Long Short-Term Memory Network	no	GC (0 or 1)
Conv2D	2D Convolutional Network	yes	R^2-score
TRGC	Time-Reversed Granger Causality	no	GC (0 or 1)

2.2.1. Temporal Causal Discovery Framework

The Temporal Causal Discovery Framework (TCDF) developed by [14] is based on the concept of a one-dimensional Temporal Convolutional Network (TCN) and is available on Github [27]. Input to the framework consists of an NxL data set consisting of N time series of equal Length L. Within the framework, one depthwise-separable TCN is used to obtain a prediction for a single source (target). The input of the network consists of the history of all time series, including the target time series. The output is the history of the target time series. An attention mechanism is added: each TCN_j has its own trainable attention vector $V_j = [vX1j, vX2j, \ldots, v_{ij}, \ldots vNj]$, that learns which of the input time series is correlated with the target by multiplying attention score v_{ij} with input time series X_i in TCN_j. When the training of the network starts, all attention scores are initialized as 1 and are, as such, adapted during training. The direction of connectivity and significance is determined using a shuffling procedure. For significance determination, one of the time series is shuffled while keeping the other one(s) intact when predicting the target. The runs with shuffled time series did not involve any model retraining. Instead, in the prediction step, the losses obtained when using the "shuffled" time series as predictors were compared with the losses obtained when using the non-shuffled time series. Only if the loss of a network increases significantly when a time series is shuffled that time series is considered a cause of the target time series. A time series X1 is only considered to be a significant contributor to another time series X2 if, in the first stage, its attention score is larger than one. Only if, after shuffling the potentially contributing time series X1, the difference between losses obtained by predicting future time steps with the unshuffled time series and losses obtained by predicting using shuffled time series is large enough, using an a priori determined threshold significance value, time series X1 is considered a significant contributor to time series X2. TCDF was run with PyTorch (version 1.4.0, www.pytorch.org, accessed on 17 December 2021).

Configuration. For TCDF, the chosen parameters were the number of hidden layers = 1, kernel size = dilation coefficient = 4 (a time-dimensional kernel), learning rate = 0.01, optimizer = Adam, number of epochs = 1000, significance threshold= 0.9998, seed = 1000. Kernel weights are initialized following a distribution with µ = 0, variance = 0.1.

2.2.2. LSTM-NUE—Long Short-Term Memory with Non-Uniform Embedding

Another connectivity measure is based on the RNN, in which directed cycling connections are present, i.e., there are feedback connections from output to input, and these connections create possibilities for memorization. A subtype of RNNs is the Long Short-Term Memory network (LSTM). This type of network provides a resolution for vanishing and exploding gradient problems in recurrent networks. It performs this by introducing gates and memory cells which also makes it very flexible towards gap length. The implementation in this study is an LSTM with Non-Uniform Embedding (NUE, a feature selection procedure) by [15], which is also publicly available [28]. NUE is an iterative selection procedure adopted from [18] to detect the most informative time steps of the predicting time series (phase one). In phase one, a vector V containing the most informative past time steps to explain the present state of a target time series X1 is obtained by iteratively adding time steps (of the time series' own past, but also of the past of the other time series) to the training set and obtaining a new model error as a time step is added. For instance, let $V = [V^{X1}{}_n, V^{X2}{}_n, V^{X3}{}_n]$ represent the vector with the most relevant past time steps to explain the present of the target time series. This selection of time steps goes on until the prediction error becomes larger than or equal to a threshold or until the maximum amount of time steps is reached. If for a certain time series X2, no time steps have been added in V, the time series is not further considered as a potential contributor to target time series X1, and it is not considered in the next phase (phase two). Phase one results in an estimation of the error variance of the full model (i.e., the model containing all relevant past time steps from different time series). In phase two, the model is fit only with this smaller set of time steps. The error of the reduced model is finally obtained by not using the values of the time series (e.g., X3) that is a potential contributor to the target time series X1. If the error ($Loss^{Reduced}$) of this reduced model is larger than the error of the full model ($Loss^{Full}$), time series X3 is considered a significant contributor to time series X1 ("X3 Granger-predicts X1").

In LSTM-NUE, no shuffling is used to determine connectivity. Instead, the significance procedure consists of two phases. Determining significance is based on (1) the selection of relevant time samples from all time series rendering a full model, after which the time series whose time samples were not selected are already as potential causes of the target time series. (2) The remaining candidates are then, as a test, subsequently excluded from the model to obtain the reduced model (i.e., the model with only the target time series as its own predictor). Hence, this exclusion phase is, to some extent, comparable with the shuffling procedure used in TCDF, given that this procedure is in this way testing the relevance of a certain time series in the prediction of another (by excluding it OR by shuffling the values).

Configuration: for LSTM-NUE, the parameters are the number of hidden layers = 1, the number of units in each layer = 30, batch size = 30, num_shift = 1, sequence_length = 20, number of epochs = 100, theta = 0.09, learning rate = 0.001, weight decay = 1×10^{-7}, min_error = 1×10^{-7} (=a priori determined error to determine whether a certain time step should be included in the final model), and train/validation split = 0.85/0.15. Default kernel initializer = "glorot_uniform", which draws samples from a uniform distribution, is used to initialize the weights of the LSTM-layer.

2.2.3. Conv2D—Two-Dimensional Convolutional Network

Finally, we propose a two-dimensional Convolutional Network (Conv2D) as a way to test whether a 2D kernel variation in TCDF has merit. The input consists of an NxL data set, which is transformed into a four-dimensional tensor (time samples of training set, window size, amount of predicting time series, 1). The source code is accessible via Github (kul-EEG-sourceconnectivity, https://github.com/irisv440/kul-EEG-sourceconnectivity, accessed on 21 September 2021).

Some important differences with TCDF are the fact that a two-dimensional kernel is used and that a cross-validation procedure, adapted for time series, is embedded in the framework. While in TCDF, a one-dimensional kernel (with height = 1) slides over the data along the time dimension (=width of the kernel, i.e., the amount of time steps considered together), in Conv2D, a two-dimensional kernel is used in which the second dimension represents the amount of time series that will be convolved together. The second dimension has an upper bound, which is the total amount of time series within the input data. We hypothesized that by adding a second dimension (feature dimension) to TCN, we could capture the most important aspects of the other time series, leading to more correct connectivity estimates. However, it was suggested (e.g., [29]) that convolving data from several time series can also cause less accurate results (in our case, this means lower Sensitivity and lower Precision) because too many time series are convolved together, possibly erasing the impact of changes in individual time series. Similar to TCDF, the input to the network consists of all time series, including the target time series. The output is a single target time series.

A second difference is cross-validation (CV) for time series. Cross-validation is a powerful method for detecting overfitting, but its implementation in time series models is not trivial, given that no leakage from future to past may exist. This issue was solved by using 6-fold cross-validation on a rolling basis based upon "TimeSeriesSplit" from the model selection module of the sklearn-library version 0.24.1 (Scikit-learn, original version released by [30]). With TimeSeriesSplit, we obtained the following train-test regime for the folds where "—"represents the unused part of the data in the corresponding fold (Figure 5).

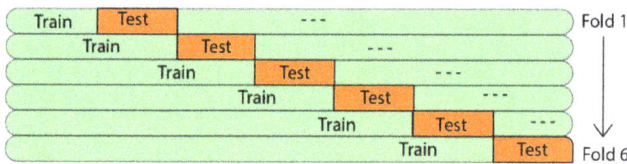

Figure 5. CV with length of first train-fold = length of first test-fold (= 1500/7).

In addition, given that connectivity may vary over longer time spans (as is also the case in Ground Truth 1), working with only one division in the train/validation/test-set (respecting past versus future) can cause false positives or false negatives since one may be training on a portion of the time series where connectivity is very strong between, for instance, X3 and X2 while validating and/or testing on a part where the same connectivity is weak (or the other way around).

As a metric for connectivity strength, the R^2-score between the real values and the predicted values of the current target is used. The better a time series pair is successful in predicting a target, the larger the similarity between the true values and the predicted values will be, hence the stronger the connectivity between the time series and target. When, for instance, two different pairs of time series X1 and X2, versus X1 and X3 are used as predictors for X1, R^2 again represents the similarity between predicted and true values of the target X1. When the prediction of X1 becomes better when predicted by time series X1 and X2 together, instead of with X1 and X3, one could conclude that connectivity is stronger between X1 and X2 than between X1 and X3. The R^2 scores themselves are obtained from the cross-validation folds, after which the average R^2 score is taken over the folds and over the number of used runs for one data set. The corresponding output is a scoring matrix representing all combinations of time series used as predictors and possible target time series. If the R^2 score is >0 and the predicting time series are considered significant (see "Connectivity Analysis using ANNs"), the obtained R^2 score can be interpreted. However, when including more than two predictors, this relationship is not so easily established anymore, given that the R^2-score still represents the connection between the target and all predicting time series together. Similar to TCDF, the direction of connectivity and significance is determined using a shuffling procedure. Significance weights are obtained by comparing training and test loss differences, after which a data-driven cutoff (here 0.70) is used to differentiate between contributing and non-contributing time series. More concretely, training difference = (first training loss)–(final training loss), where the latter is expected to be much lower than the first term, and Test difference = (first training loss)–(loss of test-indices using shuffled train data) where the latter is expected to be high because of the shuffled data; hence, one expects the test difference to be very small. Next, if the average test difference was larger than the average_training_difference * significance (=0.9998), the potential connection is considered not significant in the first place. Significance weights are obtained by (test difference/training difference). If the weight is larger than the cutoff (=0.70), the connection is considered not significant. The used significance level, as well as the cutoff for significance weights, were experimentally determined, and the final choice was based upon a data-driven approach (by experimenting with significance levels in the range of {0.70, 1} and with cutoff-scores in the range of [0.40, 0.70]). For the current kind of simulated data, these values worked well.

Configuration: for Conv2D, the parameters were as follows: number of hidden layers = 1, number of filters = 24, kernel size = {4*2, 4*3} (width*height), dilation coefficient = 1, number of epochs = 12, window size = 5, learning rate = 0.005, optimizer = "Adam", significance = 0.9998, cut-off scores for significance weights = 0.70 and number of train/test splits for CV = 6. Default kernel initializer = "glorot_uniform", was used to initialize the weights of Keras' Conv2D-layer.

2.2.4. TRGC—Time-Reversed Granger Causality

As our baseline method, Time-Reversed Granger Causality (TRGC), as implemented (by means of the Matlab function "tr_gc_test", embedded in "simulation_source_connectivity"), and evaluated by [1], was used. As stated before, the difference with "traditional" GC is the type of significance procedure. Instead of the classical way to determine significance (a likelihood ratio test), which cannot distinguish between actual versus spurious correlations due to source mixing, it determines whether the "standard" GC scores for non-reversed and reversed data have opposing directions and are both significant. In other words, direction-flipping must occur when data are time-reversed. This is referred to as conjunction-based TRGC [7]. A drawback of GC (and hence, TRGC) is that one needs to define the model order, which is feasible when the ground truth is known, such as in simulations, but in "real" EEG data, this quickly becomes a tricky problem. An advantage, on the other hand, is the fact that with TRGC, one model for all sources is constructed, after which one threshold is applied to all obtained GC scores.

Configuration. Function tr_gc_test takes as input an NxL matrix H′, the model order, the number of time steps in the time series, alpha, the type of significance test ("conservative", requiring significant GC scores with original as well as reversed data; versus a significance test based on difference scores between GC scores in normal and reversed order) and finally, the type of VAR model estimation regression mode to calculate pairwise-conditional time-domain Granger Causality scores. In this work, the model order of TRGC was set to two, we opted for "conservative" significance testing, and ordinary least squares (OLS) was used as Vector-Autoregression (VAR) estimator. We used an alpha level of 0.05, FDR corrected [31]. The corresponding p-value was taken as a threshold to binarize connectivity scores.

2.3. Performance Evaluation

The main question is whether the connections in the ground truths could be detected by the evaluated networks and by TRGC ("True Positives", TP) without detecting too many false connections ("False Positives", FP), thus connections that are not present in the ground truths. Measures based upon these are Precision, Sensitivity/Recall, and F1-score (Figure 6), which we used for comparing TCDF, LSTM-NUE, Conv2D and TRGC.

> Connectivity-related measures:
>
> 1) Sensitivity (or Recall) = $\frac{TP}{TP+FN}$
>
> 2) Precision = $\frac{TP}{TP+FP}$
>
> 3) $F1 = 2 * \frac{Precision * Recall}{Precision + Recall}$
>
> General performance measure*:
>
> Time Complexity (TC) = Runtime in seconds for five runs of one dataset on an Acer Aspire 7 A715-75G-751G, intel i7, 16 GB RAM.
>
> *Used for comparison between ANNs only.

Figure 6. Main evaluation measures.

The results on connectivity strength are not directly compared between models as they differ substantially. These strength estimates, based on the mean over five runs on the same data set, are calculated and ranked. It must be emphasized that these strength estimates are relative per model and target training series as, for each target time series, the network is trained differently. The latter implies that connection strengths obtained in the prediction of a particular Target time series X1 cannot be readily compared with connection strengths obtained in the prediction of another target time series X2. If F1 < 50%, only rankings are presented. Self-connectivity is not taken into account to avoid an overly positive perception of the results.

3. Results

3.1. Ground Truth 1

3.1.1. Connectivity Detection

In Figures 7 and 8, respectively, Sensitivity and Precision are shown per method. Some remarks, specifically with regard to Conv2D, need to be made before interpreting the results.

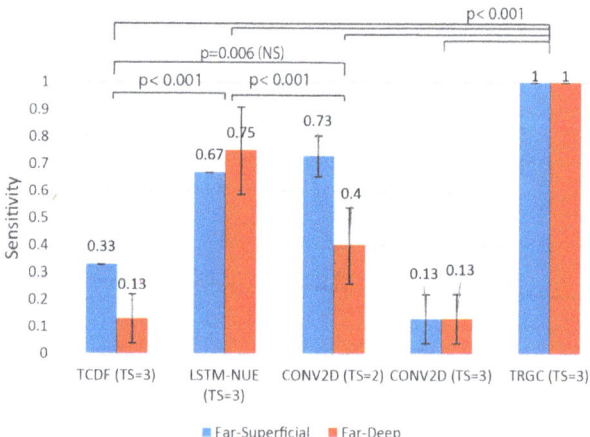

Figure 7. Mean Sensitivity for all methods (L = 1500). Sensitivity Ranking Far–Superficial: TRGC > Conv2D (TS = 2) > LSTM-NUE > TCDF > Conv2D (TS = 3). Sensitivity Ranking Far–Deep: TRGC > LSTM-NUE > Conv2D (TS = 2) > Conv2D (TS = 3) = TCDF. Abbreviation TS = amount of time series included in the predictions.

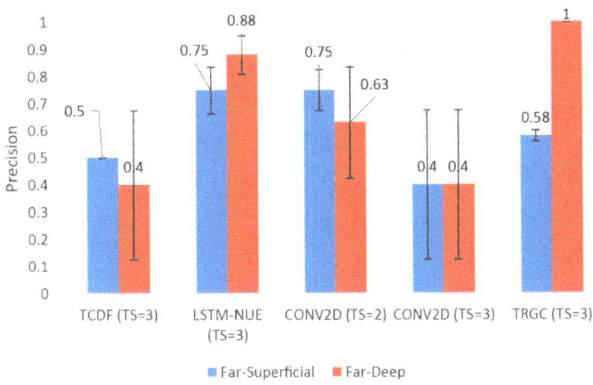

Figure 8. Mean Precision for all methods (L = 1500). Sensitivity Ranking Far–Superficial: TRGC > Conv2D (TS = 2) > LSTM-NUE > TCDF > Conv2D (TS = 3). Sensitivity Ranking Far–Deep: TRGC > LSTM-NUE > Conv2D (TS = 2) > Conv2D (TS = 3) = TCDF. Abbreviation TS = amount of time series included in the predictions.

Given that the Conv2D-model based on three predicting time series resulted in a very low Sensitivity (0.13 ± 0.18) and low Precision (0.40 ± 0.55), see Figures 7 and 8, it was not considered relevant to explore the model with three predictors further in terms of connectivity strength (for connectivity strength per ANN, see Sections 3.1.2–3.1.4). This decision was supported by the results of a Scheirer–Ray–Hare Test with model and dipole condition as factors and with follow-up Mann–Whitney U tests (Bonferroni-corrected). Superior results were obtained with Conv2D models containing two predicting time series versus three predicting time series. These results can be consulted in Appendix A (Table A1).

Hence, strength rankings are explored only with the Conv2D model with two predictors (Section 3.1.4). With regard to the model based on two predicting time series, the results obtained by looking at each predictor pair (consisting of two time series) separately revealed large differences between pairs in terms of Sensitivity and Precision. We chose to take all detected connections into account while calculating our scores instead of averaging over all predictor pairs, as it could lead to biased results. This is because if it is found that X2 predicts X1 when it is predicted together with X1 but not detected when it is predicted with X2 and X3, and the discovered connection between X2 and X1 is still included in the performance scores, this increases Sensitivity but decreases Precision. The decrease in Precision then occurs because if a false positive is found by one of the two predictor pairs, it is still counted. Option 1 was chosen to put the focus more upon detection ability and exploration. Thus, it must be kept in mind that a positive detection bias exists in all our overall two-to-one performance scores of Conv2D.

While focusing on differences in Sensitivity, the following results were obtained for the used ANNs and TRGC. A Scheirer–Ray–Hare Test with model and dipole condition as factors revealed no statistically significant interaction (using alpha = 0.05) between the effects of the type of connectivity method and dipole condition (p = 0.63), nor the main effect of dipole condition itself (p = 0.63). However, a simple main effects analysis showed that the type of connectivity method does have a statistically significant effect on Sensitivity (H (4,40) = 38,159, p < 0.001). Follow-up two-sided Mann–Whitney U tests (Bonferroni-corrected: alpha = 0.05, alpha adjusted = 0.005), carried out across dipole conditions, show significant and marginally significant differences between the following methods. Median scores (denoted as Mdn) are reported. In contrast to TCDF (Mdn = 0.33), smaller contributions of one time series to another could be detected with LSTM-NUE (Mdn = 0.67), p < 0.001. The difference between TCDF and Conv2D (Mdn = 0.67) with two time series as predictors was only marginally significant after correcting for multiple comparisons, $p = 0.006$. TRGC ($Mdn = 1$), however, outperformed all ANN models in terms of Sensitivity (p-values denoting differences with all other methods < 0.001). Finally, while no significant difference was found between LSTM-NUE ($Mdn = 0.67$) and Conv2D ($Mdn = 0.67$) with two time series as predictors ($p = 0.239$), LSTM-NUE performed significantly better than Conv2D with three time series as predictors ($Mdn = 0$), $p < 0.001$. Rankings are described below in Figure 7 to provide qualitative comparisons. Note that in Figure 7, mean scores M for each dipole condition is still reported, given that the current results were obtained with small sample sizes. Hence, differences between dipole conditions may still appear once statistical power is increased (i.e., by using more data sets) and given that differences between dipole conditions were, to some extent, expected.

With regard to Precision, the results of a Scheirer–Ray–Hare Test with method and dipole condition as factors were not significant, albeit a marginally significant result for method ($H (4,40) = 8.79$, $p = 0.07$) was obtained. Hence, no follow-up tests were carried out.

Thus, we rely upon rankings only for our qualitative description (in terms of mean scores M, taking dipole condition into account) of the data. In the Far–Superficial dipole condition, Conv2D with TS = 2 and LSTM-NUE obtain both a Precision of $M = 0.75$ (± 0.15, 0.17, respectively), followed by TRGC ($M = 0.58 \pm 0.05$) and TCDF ($M = 0.50 \pm 0$). Precision is lowest in Conv2D with TS = 3 ($M = 0.40 \pm 0.55$). However, in the Far–Deep dipole condition, TRGC obtains perfect Precision ($M = 1 \pm 0$), followed by LSTM-NUE ($M = 0.88 \pm 0.14$) and Conv2D with TS = 2 ($M = 0.63 \pm 0.41$). The qualitatively lower Precision score of TRGC in the Far–Superficial condition turned out to be mainly due to two consistently observed false-positive connections that were not detected in the Far–Deep dipole condition.

When summarizing the results in terms of F1-scores, the following ranking was obtained for the Far–Superficial condition: TRGC ($M = 0.73 \pm 0.04$) = Conv2D with TS = 2 ($M = 0.73 \pm 0.09$) > LSTM-NUE ($M = 0.70 \pm 0.07$) > TCDF ($M = 0.40 \pm 0.0$) > Conv2D with TS = 3 ($M = 0.20 \pm 0.27$).

For the Far–Deep condition, the F1-score ranking was as follows: 1 ± 0 (TRGC) > 0.75 ± 0.17 (LSTM-NUE) > 0.47 ± 0.31 (Conv2D with TS = 2) > 0.20 ± 0.27 (Conv2D with TS = 3) = 0.20 ± 0.27 (TCDF).

3.1.2. TCDF

With regard to TCDF (Figure 9), only two mean attention scores were significant, and solely in the Far–Superficial dipole condition and for two different targets (X2 and X3), such that a target-wise comparison cannot be made. More concretely, a connection X3 → X2 was found (in accordance with the ground truth), as well as a connection X2 → X3 (unlike the ground truth).

	Far-Superficial			Far-Deep			Truth		
	T X1	T X2	T X3	T X1	T X2	T X3	T X1	T X2	T X3
X1	-	NS	NS	-	NS	NS	-	3	NS
X2	NS	-	1.07	NS	-	NS	2	-	NS
X3	NS	1.43	-	NS	NS	-	NS	1	-

Figure 9. TCDF, Mean attention score rankings. Colors: "1", green, denoting the highest attention score in one TCN—one column). The columns represent the targets (T X1, T X2, T X3), the rows the predictors (X1, X2, X3). Self-connectivity is excluded.

3.1.3. LSTM-NUE

Next, when focusing on LSTM-NUE, as can be seen from the colors from Figure 10, for target time series X3, the GC scores (in both dipole conditions) were higher than expected according to the ground truth. Unexpected GC scores are surrounded by black rectangles in the top panel. Column-wise strength rankings (rankings for one particular target) are correct for two out of three targets (X1, X2) in both conditions, as can be seen by comparing with column-wise Ground Truth 1 (Figure 10, bottom right panel). The overall ranking in the Far–Deep condition was more in accordance with the overall ranking in Ground Truth 1 (Figure 10, bottom left panel) than the ranking found in the Far–Superficial condition because connectivity strength was observed to be the weakest for the corresponding false positives (as shown in yellow in the top panel).

	Far-Superficial			Far-Deep		
	T X1	T X2	T X3	T X1	T X2	T X3
X1	-	5.50E-03 (± 1.10E-02)	NS	-	1.09E-02 (± 1.26E-02)	2.34E-03 (± 4.69E-03)
X2	3.78E-02 (± 1.48E-02)	-	3.20E-02 (± 3.32E-02)	3.08E-02 (± 3.66E-02)	-	2.75E-03 (± 5.51E-03)
X3	NS	1.19E-02 (± 8.98E-03)	-	NS	1.11E-01 (± 6.37E-02)	-

	Truth 1: Overall ranking			Truth 1: Ranking per Column		
	T X1	T X2	T X3	T X1	T X2	T X3
X1	-	3	NS	-	2	NS
X2	2	-	NS	1	-	NS
X3	4	1	-	2	1	-

Figure 10. LSTM-NUE, Neural-Network Granger Scores rankings (Top) versus Truth 1 (Bottom), excluding self-connectivity. Color coding: dark green > light green > yellow > orange > red. The columns represent the targets, the rows the time series used for prediction

3.1.4. Conv2D

In Figure 11, R^2-strength rankings for Conv2D with two time series are shown. Given that for Conv2D, adding a third time series did not work out well, only rankings per predictor pair could be obtained. When R^2-strength is shown in the upper two panels of Figure 11, it means that the current time series pair is a significant contributor. Significance weights, which denote significant contributions of one time series to a target time series (instead of R^2 scores denoting a connection between a certain pair of predicting time series and one target time series), are reported between brackets. They were obtained as described in Section 2.2.3 and were considered significant if the cutoff of 0.70 was not exceeded. The lower two panels show a ranking (with 1 being the most active connection and 4 the least active connection).

Figure 11. Conv2D, R^2-strength score for time series pairs (Top) versus ranking of connections in the Ground Truth 1 (Bottom, 1 being the strongest), including self-connectivity. Color coding: dark green > light green > yellow > orange > red. The columns represent the targets, the rows the time series pairs used for predicting the target time series.

It can be seen from Figure 11 that, in the prediction of Target X1, out of three direct connections, only one is not significant (i.e., X2, X3 → X1, Top-row), but this is only the case when Target X1 is not included as a predictor. Regarding an overall ranking (Ground Truth 1, Bottom-Left), it can be seen that X3 has the strongest self-connectivity, while for X1 and X2, self-connectivity is almost the same. This is observed in our results as well ($R^2 = 0.36, 0.35$ in both dipole conditions). Moreover, the obtained R^2 strength scores are not, or barely, dependent on the dipole condition. Next, while inspecting these results column-wise (hence, target-wise), a stronger connection between predictors X1, X2 and Target X1 than between predictors X1, X3 and target X1 were expected. However, these connections are quite similar ($R^2 = 0.35$ versus $R^2 = 0.34$ in the Far–Superficial condition, $R^2 = 0.36$ versus $R^2 = 0.34$ in the Far–Deep condition). While predicting target X2 using X1, X3, a significant, correct contribution from X1 to X2 is found (significance weight = 0.232, 0.120, Far–Superficial and Far–Deep condition, respectively), as well as a correct contribution from X3 to X2 (significance weight = 0.001, 0.048, Far–Superficial and Far–Deep condition, respectively). However, connectivity strength R^2 is very low ($R^2 = 0.02$ in both dipole conditions) in comparison with the situation in which target X2 is included in the predictor pair and in which case X3 is also considered a significant contributor (predictor pair = X2, X3, $R^2 = 0.36, 0.36$, significance weights = 0.341, 0.210 for X3, Far–Superficial and Far–Deep condition, respectively). The ranking for Target X2 is correct, as was the ranking for X1. Finally, we expected similar rankings for X2; X3 predicting X3 as for X1; X3 predicting X3 since neither X1 nor X2 contribute to X3. This is indeed the case for both conditions. As expected, significance weights for individual contributions of X1 and X2 to X3 were not significant (significance weights >0.70 in both dipole conditions).

3.1.5. Time Complexity

Finally, we assessed runtimes in seconds for one data set (including averaging over five runs) w.r.t. the training of the ANNs. Runtimes with three time series as predictors (TS = 3), Length L = 1500, dipole condition = Far–Superficial are shown in Figure 12, as well as the runtime of Conv2D with two time series as predictors. The runtime of Conv2D with only two time series as predictors was 1048 s. All runs were performed with an Acer Aspire 7 A715-75G-751G, intel i7, 16 GB RAM.

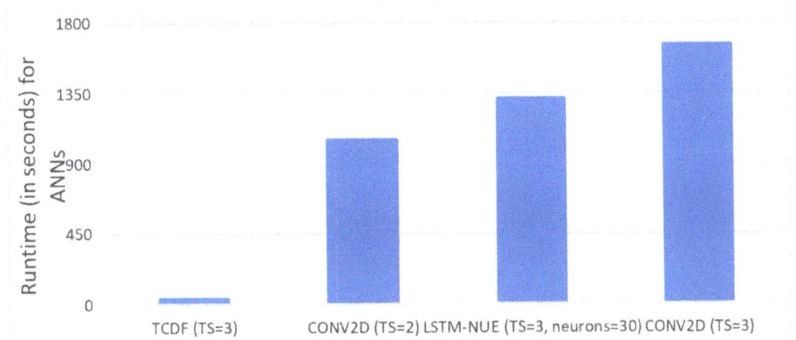

Figure 12. Runtime for the training of all ANNs, with 3 time series (TS = 3), L = 1500. An extra comparison showing the runtime of Conv2D with two time series as predictors (TS = 2) is shown (but all datasets contain 3 time series). "Neurons" = the number of hidden layer neurons.

3.2. Ground Truth 2

With only one true connection and excluding self-connectivity, it was found that none of the methods, except for LSTM-NUE and TRGC (LSTM-NUE, TRGC: Sensitivity $M = 1 \pm 0$ in both dipole conditions), were able to detect this connection in none of the runs or datasets (Table 2).

Table 2. Scores of the ANNs in comparison with TRGC, using Ground Truth 2. Results are based upon datasets where all 3 time series (TS) were included as predictors, with one exception: results from Conv2D with two time series as predictors, indicated with *, were also included.

	Far-Superficial			Far-Deep		
	Sensitivity	Precision	F1	Sensitivity	Precision	F1
TRGC	1 ± 0.00	0.71 ± 0.29	0.79 ± 0.21	1 ± 0.00	0.70 ± 0.29	0.79 ± 0.21
TCDF	0 ± 0.0	0 ± 0.0	0 ± 0	0 ± 0.0	0 ± 0.0	0 ± 0
LSTM-NEU	1 ± 0.0	0.37 ± 0.13	0.53 ± 0.14	1 ± 0	0.43 ± 0.09	0.60 ± 0.09
CONV2D * (TS=2)	0 ± 0.0	0 ± 0.0	0 ± 0.0	0.0 ± 0.0	0.0 ± 0.0	0.0 ± 0.0
Conv2D (TS=3)	0 ± 0.0	0 ± 0.0	0 ± 0.0	0 ± 0.0	0 ± 0.0	0 ± 0.0

The results of a Scheirer–Ray–Hare Test with method and dipole condition as factors reveal, as expected, a main effect of connectivity method on Sensitivity ($H(4,40) = 47.66$, $p < 0.001$) as well as Precision ($H(4,40) = 48.10$, $p < 0.001$). The interaction between method and dipole condition, nor dipole condition itself were significant (Sensitivity: $H(4,40) = 0.05$, $p = 0.99$, $H(1,40) = 0.01$, $p = 0.91$, interaction and dipole condition effect, respectively; Precision: $H(4,40) = 0.03$, $p = 0.99$, $H(1,40) = 0.00$, $p = 0.95$, interaction and dipole condition effect, respectively). Looking into the effects of different connectivity methods using follow-up Mann–Whitney U tests (Bonferroni-corrected: alpha = 0.05, alpha adjusted = 0.005), significant differences in Sensitivity and Precision between LSTM-NUE/TRGC versus all other methods were found ($p < 0.005$). No differences in Sensitivity between

TRGC (*Mdn* = 1) and LSTM-NUE (*Mdn* = 1) were found (p = 0.21), while Precision was significantly higher for TRGC (*Mdn* = 0.67) than for LSTM-NUE (*Mdn* = 0.42), $p < 0.001$. It is not surprising that comparisons of any other ANN method than LSTM-NUE with TRGC were significant, given that these methods had a Sensitivity and Precision of zero. Even though dipole condition did not turn out to exhibit a significant effect on Sensitivity nor Precision in our data sets, this distinction remains theoretically important. We summarized the qualitative differences below.

Sensitivity and Precision were in both dipole conditions 0 while using TCDF and using two different configurations of Conv2D (once with two time series as predictors, once with three time series as predictors). In contrast, Precision was M = 0.37 (± 0.13) for LSTM-NUE while TRGC obtained a precision of M = 0.71 (± 0.29) in the Far–Superficial dipole condition. In the Far–Deep condition, performance of TRGC remained almost the same (Sensitivity $M = 1 \pm 0$, Precision $M = 0.70 \pm 0.29$) while it became slightly higher (in contrast to the Far–Superficial condition) for LSTM-NUE (Sensitivity $M = 1 \pm 0$, Precision $M = 0.43 \pm 0.09$). F1-scores were $M = 0.79 \pm 0.21$ and $M = 0.53 \pm 0.14$ for TRGC and LSTM-NUE, respectively, in the Far–Superficial condition and $M = 0.79 \pm 0.21$, $M = 0.60 \pm 0.09$ in the Far–Deep condition (while being zero for all the other ANNs).

4. Discussion

While considering Sensitivity and Precision, it was shown that, among the ANNs, LSTM-NUE yielded superior results in terms of Sensitivity, resulting in statistically significant differences with the other ANNs except for Conv2D with TS = 2. In terms of Precision, however, no significant differences among the ANNs were found while using Ground Truth 1. TRGC outperformed all ANNs in terms of Sensitivity but, statistically, no differences in Precision were found given that the main effect of the connectivity method was only marginally significant. The lack of a statistically significant effect of connectivity method on Precision, as well as the lack of an effect of dipole condition, and the lack of an interaction effect on both Sensitivity as well as Precision are quite counterintuitive. Indeed, given (1) the patterns observed across both Ground Truths and (2) the results from [1], which convincingly showed effects of different dipole conditions on connectivity patterns as well as interaction effects of connectivity method and dipole condition, one could at least expect an effect of dipole condition. For instance, in [1], it was shown that with an SNR of 0.9 and in a Far–Superficial dipole condition, false positives (as related to Precision) were rather rare, while for other dipole conditions, the percentage of false positives increases (hence decreasing Precision). A related (solely qualitative) observation is the variability in the results of the ANNs (as became obvious through the standard deviations from the mean as depicted in Figures 7 and 8) versus the stability of results produced by TRGC. In particular, ANNS seems to exhibit an increased variability in performance in the Far–Deep Condition (in contrast to the Far–Superficial condition), while almost no such variability is observed for TRGC. A possible culprit could be the initial randomization of the weights in ANNs, but how this instability could differ between architectures or between dipole conditions is unclear and deserves attention in future studies. One of the most important observations of Ground Truth 1 is the relatively poor Precision score of TRGC in the Far–Superficial condition, albeit that a difference with the Far–Deep dipole condition could not be statistically confirmed. More data may be needed to confirm the observed trends. The above-mentioned contrasting results are further discussed below, together with possible explanations with regard to the used connectivity methods.

Using Ground Truth 2, no differences in Sensitivity between TRGC and LSTM-NUE were found given that both methods returned almost always a Sensitivity of one, while Precision was significantly higher for TRGC than for LSTM-NUE. The other ANNs did not detect any connection. The good performance of TRGC regarding Precision is not surprising. In [1], it was already shown that TRGC outperformed Multivariate Granger Causality (MVGC), especially when it comes to false positives (as reflected in a lower False Positive Rate), which is logical given that the introduction of time-reversal could indeed allow for a better distinction between correlated time series (due to linear mixtures of EEG signals) and true temporal precedence of one time series with regard to another. Although the idea of TRGC is relatively new (as it was first proposed in 2013, by [8]) in comparison to, for instance, bivariate GC and MVGC, due to its appealing theoretical properties as well as its further validation by [7], it was quickly picked up in the field, given its relevance for, among others, EEG source connectivity. Recent developments include, for instance, variations in TRGC that allow for other than normal distributions [32].

In summary, it became clear that, among the ANNs, LSTM-NUE obtained better Sensitivity scores and (although only statistically confirmed using Ground Truth 2) better Precision scores. TRGC outperformed the ANNs in terms of Sensitivity, but in the case of Ground Truth 1, questions arose surrounding its Precision in the Far–Superficial dipole condition (although its Precision was significantly better in Ground Truth 2, without any indication of possible differences between dipole conditions). While all connections were discovered, two false positives were detected relatively consistently, indicating that even with time-reversal there is, in certain circumstances, an over-detection of connections. The lack of performance of TCDF and Conv2D in Ground Truth 2 cannot be due to the location of the two fixed dipoles since they were located at the exact same location as in Ground Truth 1. Hence, we suspect that the moving nature of the sending dipole explains (at least partly) the lack of Sensitivity in TCDF and Conv2D. Taking the results from both Ground Truths together, both LSTM-NUE and TRGC are clearly more sensitive, but they both still tend towards over-detection.

With regard to the score strength rankings, not much can be said about TCDF given that the mean attention scores were significant only for two time series in the Far–Superficial dipole condition, from which one was a falsely detected connectivity (i.e., a false positive). In contrast to TCDF, with LSTM-NUE, for two out of three targets, correct column-wise rankings were obtained for Ground Truth 1. For Conv2D (with TS = 2), correct rankings for predictor pair were found in terms of R^2-scores, also for two out of three targets. When looking closer to the contributions of individual time series, it was found that predicting, for instance, X1, with itself and another time series works better than predicting it without the past of X1, which is logical. The fact that adding more predictors (i.e., Conv2D with TS = 3) did not work out is obviously the most problematic aspect of Conv2D. Once a third predictor was added, performance dropped substantially, and it was hypothesized that this could be due to the fact that it was convolving rather uncorrelated or only slightly correlated time series together confuses the two-dimensional network to the extent that no proper prediction can be made. The fact that channels are not kept separate such as in a depthwise-separable architecture, may play an important role in this aspect. Finally, with regard to runtimes (time needed to train a model), LSTM-NUE was together with Conv2D, TS = 3 the most time-consuming method, which calls for a trade-off between accuracy and Time Complexity. It is especially the non-uniform embedding strategy (NUE) that is responsible for the high Time Complexity. However, in [15], it was shown that the current LSTM-model could also produce reasonable results without implementation of the NUE strategy, thereby lowering its Time Complexity drastically.

Moreover, in [15], it was shown that LSTM-NUE could cope with different types of ground truths (linear, non-linear and non-linear with varying length lags), as confirmed in our work. Contrary to [15], we, in addition, had Ground Truth 2 with a moving dipole (i.e., the "Sender") which worked relatively well for LSTM-NUE. Hence, the latter can cope not only with time-varying parameters but also, to some extent, with changing dipole locations. Both TCDF and Conv2D cope far less well with a moving sender, probably (or at least partly) because of the occurrence of both closeness and deepness in the same setting, which has an impact on how signals are transformed by source reconstruction. TCDF and Conv2D are, in contrast to LSTM-NUE, not a part of the family of Recurrent Neural Networks and therefore do not contain feedback loops. The LSTM is particularly known for its excellent memory properties by virtue of its gates that help to remember versus forget certain time samples. In general, the better memory properties of an LSTM in combination with the NUE approach probably play an important role in dealing with variations over time. An LSTM may also be better in looking through (uncorrelated) noise components because it remembers formerly seen time samples better and, subsequently, should be better in detecting (even weak) patterns over time, also when occluded by noise. This, in turn, may make it easier to deal with more challenging dipole locations or with heavier data transformations. However, this same property could also make an LSTM more sensitive to correlated noise from source mixing. TCDF, on the other hand, has the advantage of a very low Time Complexity, at least partly due to its sparsity in interconnection weights (given its depthwise-separable architecture), but it seems less able to distinguish correlation from causation. This may be due to the lack of feedback loops, an "active" memory feature that makes it difficult to distinguish true patterns from noise over longer time intervals. In this study, TCDF was tuned as such that not too many false positives were detected (given its problem of distinguishing correlation from causation), and this more "conservative" configuration may have led to its low Sensitivity. Overall, we can conclude that, among the ANN models, LSTM-NUE performed best in terms of Sensitivity and Precision regardless of which ground truth was used even though no shuffling or time-reversal was used for connectivity assessment. The contrasting results of TRGC in terms of Precision between dipole conditions in Ground Truths 1 and 2 are puzzling and clearly show an "oversensitivity" of TRGC under certain circumstances. Still, TRGC and LSTM-NUE yielded acceptable-to-good results, albeit both suffer from over-detection. An interesting new finding is the fact that an LSTM is, to some extent, able to provide an answer to the question of whether connectivity between sources is present or absent, at least for source-reconstructed, simulated EEG data. The fact that too many faulty connections were detected (especially in Ground Truth 2) calls for improvements. One possibility is to use LSTM-NUE as part of a masking approach, on top of which another learner is stacked. This masking approach has already led to many advantages in source localization [25], and it may also facilitate connectivity detection with ANNs, especially when overly sensitive to it. In this sense, other ANNs, even with a lower Time Complexity than that of LSTM-NUE, could possibly also be considered as potentially directed connectivity estimators.

An obvious future step is testing whether ANNs can also be applied to real EEG data, albeit that several possible caveats should be taken into account. First and foremost, as shown by [1], under low noise conditions, dipole conditions may matter less, but differences between dipole conditions could become more obvious (i.e., more disturbing) under higher noise levels. Even long-established connectivity methods suffer from this. Since controlling noise levels is hard, reasonably one could opt for EEG-data for which (1) the contributing brain areas are rather superficially located, (2) the connectivity patterns are relatively well known and preferably supported by both high-density EEG and fMRI-data so that a performance evaluation becomes feasible since no ground truth is available for real EEG-data. Testing ANNs and contrasting them with TRGC/other established methods using vision-related or motor-related EEG-datasets makes thus more sense than testing them with data with relatively unknown connectivity patterns. Regions of Interest (ROIs) can be defined based upon previous knowledge about involved brain areas. As for source localization, a reasonable choice is eLORETA. Data-driven approaches (as opposed to ROI-selection), e.g., data-driven clustering [33], seem only reasonable in a later stage when the value of the used ANN is proven on real EEG data.

5. Conclusions

Some types of neural networks, in particular LSTMs, may be considered for estimating the directed connectivity of reconstructed EEG Sources. However, no method is flawless, and we showed that even an established method such as TRGC can generate faulty estimates. This calls for further developments. There is much potential for a hybrid approach, in which a neural network could be used as a preprocessing step to chart the interesting directed connectivity patterns, after which a conventional method is applied for estimating them.

Author Contributions: Conceptualization, M.M.V.H., A.F. and I.V.; methodology, I.V. and A.F.; formal analysis, I.V.; resources (scripts), A.F. and I.V.; original draft preparation, I.V.; writing, review and editing, A.F and M.M.V.H.; visualization, I.V.; supervision, A.F. and M.M.V.H.; project administration, A.F. All authors have read and agreed to the published version of the manuscript.

Funding: I.V. performed this work as part of her Master in Artificial Intelligence thesis at KU Leuven. A.F. is supported by a grant from the Belgian Fund for Scientific Research—Flanders (FWO 1157019N). M.M.V.H. is supported by research grants received from the European Union's Horizon 2020 research and innovation programme under grant agreement No. 857375, the special research fund of the KU Leuven (C24/18/098), the Belgian Fund for Scientific Research—Flanders (G0A4118N, G0A4321N, G0C1522N), and the Hercules Foundation (AKUL 043).

Informed Consent Statement: Not applicable.

Data Availability Statement: Not applicable.

Conflicts of Interest: The authors declare no conflict of interest. The funders had no role in the design of the study; in the collection, analyses, or interpretation of data; in the writing of the manuscript, or in the decision to publish the results.

Appendix A

Table A1. Scheirer–Ray–Hare Test using Sensitivity as a criterion. Alpha level = 0.05. Significant results shown in bold. Models: Conv2D, TS = 2, kernel size 4×2 versus 4×4, Conv2D, TS = 3, kernel size 4×3 versus 4×4, with TS denoting the amount of time series. Model (H (3,32) = 16.04, $p = 0.001$) and dipole condition (H (1,32) = 4.53, $p = 0.03$) have significant effects on Sensitivity. Post hoc Mann–Whitney U tests (Bonferroni-corrected alpha level = 0.008) revealed that Sensitivity was significantly higher for Conv2D, TS = 2 (Mdn = 0.67, 0.67), versus Conv2D, TS = 3 (Mdn = 0.00, 0.17), $p = 0.006$, $p = 0.007$ for differing kernel sizes.

Predictor	Sum of Squares	df	Mean Square	H	p-Value
Dipole condition	577.60	1		4.53	0.033
Conv2D model	2045.35	3		16.04	0.001
Interaction	350.15	3		2.75	0.432
Within	1999.40	32			
Total	4972.50	39	127.5		

References

1. Anzolin, A.; Presti, P.; Van De Steen, F.; Astolfi, L.; Haufe, S.; Marinazzo, D. Quantifying the effect of demixing approaches on directed connectivity estimated between reconstructed EEG sources. *Brain Topogr.* **2019**, *32*, 655–674. [CrossRef] [PubMed]
2. Baccalá, L.A.; Sameshima, K. Partial directed coherence: A new concept in neural structure determination. *Biol. Cybern.* **2001**, *84*, 463–474. [CrossRef] [PubMed]
3. Friston, K.J.; Harrison, L.; Penny, W. Dynamic causal modelling. *Neuroimage* **2003**, *19*, 1273–1302. [CrossRef]
4. Büchel, C.; Friston, K.J. Modulation of connectivity in visual pathways by attention: Cortical interactions evaluated with structural equation modelling and fMRI. *Cereb. Cortex* **1997**, *7*, 768–778. [CrossRef]
5. Bernasconi, C.; König, P. On the directionality of cortical interactions studied by structural analysis of electrophysiological recordings. *Biol. Cybern.* **1999**, *81*, 199–210. [CrossRef]
6. Tank, A.; Covert, I.; Foti, N.; Shojaie, A.; Fox, E.B. Neural granger causality. *IEEE Trans. Pattern Anal. Mach. Intell.* **2021**. [CrossRef]
7. Winkler, I.; Panknin, D.; Bartz, D.; Müller, K.R.; Haufe, S. Validity of time reversal for testing Granger causality. *IEEE Trans. Signal Process.* **2016**, *64*, 2746–2760. [CrossRef]
8. Haufe, S.; Nikulin, V.V.; Müller, K.R.; Nolte, G. A critical assessment of connectivity measures for EEG data: A simulation study. *Neuroimage* **2013**, *64*, 120–133. [CrossRef]
9. Maier, H.; Dandy, G. Artificial neural networks: A flexible approach to modelling. *Water* **2004**, *31*, 55–65.
10. Leijnen, S.; van Veen, F. The neural network zoo. *Proceedings* **2020**, *47*, 9. [CrossRef]
11. Chung, J.; Gulcehre, C.; Cho, K.; Bengio, Y. Empirical evaluation of gated recurrent neural networks on sequence modeling. *arXiv* **2014**, arXiv:1412.3555.

12. Gers, F.A.; Schraudolph, N.N.; Schmidhuber, J. Learning precise timing with LSTM recurrent networks. *J. Mach. Learn. Res.* **2002**, *3*, 115–143.
13. He, Y.; Zhao, J. Temporal convolutional networks for anomaly detection in time series. *J. Phys. Conf. Ser.* **2019**, *1213*, 042050. [CrossRef]
14. Nauta, M.; Bucur, D.; Seifert, C. Causal discovery with attention-based convolutional neural networks. *Mach. Learn. Knowl. Extr.* **2019**, *1*, 312–340. [CrossRef]
15. Wang, Y.; Lin, K.; Qi, Y.; Lian, Q.; Feng, S.; Wu, Z.; Pan, G. Estimating brain connectivity with varying-length time lags using a recurrent neural network. *IEEE Trans. Biomed. Eng.* **2018**, *65*, 1953–1963. [CrossRef]
16. Wan, R.; Mei, S.; Wang, J.; Liu, M.; Yang, F. Multivariate temporal convolutional network: A deep neural networks approach for multivariate time series forecasting. *Electronics* **2019**, *8*, 876. [CrossRef]
17. Bai, S.; Kolter, J.Z.; Koltun, V. An empirical evaluation of generic convolutional and recurrent networks for sequence modeling. *arXiv* **2018**, arXiv:1803.01271.
18. Montalto, A.; Stramaglia, S.; Faes, L.; Tessitore, G.; Prevete, R.; Marinazzo, D. Neural networks with non-uniform embedding and explicit validation phase to assess Granger causality. *Neural Netw.* **2015**, *71*, 159–171. [CrossRef]
19. Lai, M.; Demuru, M.; Hillebrand, A.; Fraschini, M. A comparison between scalp- and source-reconstructed EEG networks. *Sci. Rep.* **2018**, *8*, 12269. [CrossRef]
20. Omidvarnia, A.; Mesbah, M.; O'Toole, J.M.; Colditz, P.; Boashash, B. Analysis of the time-varying cortical neural connectivity in the newborn EEG: A time-frequency approach. In Proceedings of the International Workshop on Systems, Signal Processing and their Applications (WOSSPA), Tipaza, Algeria, 9–11 May 2011; pp. 79–182.
21. Fahimi Hnazaee, M.; Khachatryan, E.; Chehrazad, S.; Kotarcic, A.; De Letter, M.; Van Hulle, M.M. Overlapping connectivity patterns during semantic processing of abstract and concrete words revealed with multivariate Granger Causality analysis. *Sci. Rep.* **2020**, *10*, 2803. [CrossRef]
22. Winterhalder, M.; Schelter, B.; Hesse, W.; Schwab, K.; Leistritz, L.; Klan, D.; Bauer, R.; Timmer, J.; Witte, H. Comparison of linear signal processing techniques to infer directed interactions in multivariate neural systems. *Signal Process.* **2005**, *85*, 2137–2160. [CrossRef]
23. Huang, Y.; Parra, L.C.; Haufe, S. The New York Head-A precise standardized volume conductor model for EEG source localization and tES targeting. *Neuroimage* **2016**, *140*, 150–162. [CrossRef] [PubMed]
24. Pascual-Marqui, R.D. Discrete, 3D distributed, linear imaging methods of electric neuronal activity. Part 1: Exact, zero error localization. *arXiv* **2007**, arXiv:0710.3341.
25. Faes, A.; de Borman, A.; Van Hulle, M.M. Source space reduction for eLORETA. *J. Neural Eng.* **2021**, *18*, 066014. [CrossRef]
26. Pascual-Marqui, R.D.; Lehmann, D.; Koukkou, M.; Kochi, K.; Anderer, P.; Saletu, B.; Tanaka, H.; Hirata, K.; John, E.R.; Prichep, L.; et al. Assessing interactions in the brain with exact low-resolution electromagnetic tomography. *Philos. Trans. A Math. Phys. Eng. Sci.* **2011**, *369*, 3768–3784. [CrossRef]
27. Nauta, M. Temporal Causal Discovery Framework. Available online: https://github.com/M-Nauta/TCDF (accessed on 22 September 2021).
28. Feng, S. RNN-GC. Available online: https://github.com/shaozhefeng/RNN-GC (accessed on 15 July 2021).
29. Wales, J. Time Series Forecasting with 2D Convolutions. Available online: https://towardsdatascience.com/time-series-forecasting-with-2d-convolutions-4f1a0f33dff6 (accessed on 15 July 2021).
30. Pedregosa, F.; Varoquaux, G.; Gramfort, A.; Michel, V.; Thirion, B.; Grisel, O.; Blondel, M.; Prettenhofer, P.; Weiss, R.; Dubourg, V.; et al. Scikit-learn: Machine learning in Python. *J. Mach. Learn. Res.* **2011**, *12*, 2825–2830.
31. Benjamini, Y.; Hochberg, Y. Controlling the false discovery rate: A practical and powerful approach to multiple testing. *J. R. Stat. Soc. Ser. B. Stat. Methodol.* **1995**, *57*, 289–300. [CrossRef]
32. Chvosteková, M.; Jakubík, J.; Krakovská, A. Granger Causality on forward and Reversed Time Series. *Entropy* **2021**, *23*, 409. [CrossRef]
33. Wang, S.H.; Lobier, M.; Siebenhühner, F.; Puoliväli, T.; Palva, S.; Palva, J.M. Hyperedge bundling: A practical solution to spurious interactions in MEG/EEG source connectivity analyses. *Neuroimage* **2018**, *173*, 610–622. [CrossRef]

Article

Structure-Function Coupling Reveals Seizure Onset Connectivity Patterns

Christina Maher [1,2,*], Arkiev D'Souza [2,3], Michael Barnett [2,4,5], Omid Kavehei [1], Chenyu Wang [2,3,5] and Armin Nikpour [4,6]

1 School of Biomedical Engineering, Faculty of Engineering, The University of Sydney, Darlington, NSW 2008, Australia
2 Brain and Mind Centre, The University of Sydney, Camperdown, NSW 2050, Australia
3 Translational Research Collective, Faculty of Medicine and Health, The University of Sydney, Camperdown, NSW 2050, Australia
4 Department of Neurology, Royal Prince Alfred Hospital, Camperdown, NSW 2050, Australia
5 Sydney Neuroimaging Analysis Centre, Camperdown, NSW 2050, Australia
6 Central Clinical School, Faculty of Medicine and Health, The University of Sydney, Camperdown, NSW 2050, Australia
* Correspondence: christina.maher@sydney.edu.au; Tel.: +61-02-9114-4187

Abstract: The implications of combining structural and functional connectivity to quantify the most active brain regions in seizure onset remain unclear. This study tested a new model that may facilitate the incorporation of diffusion MRI (dMRI) in clinical practice. We obtained structural connectomes from dMRI and functional connectomes from electroencephalography (EEG) to assess whether high structure-function coupling corresponded with the seizure onset region. We mapped individual electrodes to their nearest cortical region to allow for a one-to-one comparison between the structural and functional connectomes. A seizure laterality score and expected onset zone were defined. The patients with well-lateralised seizures revealed high structure-function coupling consistent with the seizure onset zone. However, a lower seizure lateralisation score translated to reduced alignment between the high structure-function coupling regions and the seizure onset zone. We illustrate that dMRI, in combination with EEG, can improve the identification of the seizure onset zone. Our model may be valuable in enhancing ultra-long-term monitoring by indicating optimal, individualised electrode placement.

Keywords: focal epilepsy; diffusion imaging; electroencephalography; structure-function coupling; seizure onset; structural connectivity; functional connectivity

1. Introduction

The assessment of patients with focal epilepsy using a combination of structural data derived from diffusion MRI (dMRI) and functional data from electroencephalography (EEG) is gaining increased appeal [1–3]. In the brain, structural connectivity refers to an anatomical link between two or more brain regions. Connnectomes generated from diffusion MRI, can represent the strength of structural connectivity between specific brain regions. Functional connectivity is inferred from the spatio-temporal relationship between electrophysiological signals from two or more structurally discrete regions [4]. Structural connectivity is believed to give rise to functional and network behaviour [5]. In a mechanistic sense, the composition of white matter can be expected to influence the flow of activity and connectivity between neuronal populations. Therefore, if EEG functions as a tool to observe the flow of activity, the connectivity measurements from EEG can be presumed to closely resemble connectivity measurements from structural MRI. In epilepsy, structure-function coupling is proposed to have a role in identifying seizure propagation patterns [6,7], seizure generalisation [1] and predicting post-surgery seizure freedom [8,9]. Diffusion MRI derived tractography,

in conjunction with EEG, can enable the quantification of structural connectivity between different brain regions. However, in epilepsy, dMRI is held to be in the experimental realm [2]. Therefore, though there is consensus on the significance of structural connectivity information in patient diagnosis, the utility of dMRI as a routine clinical test has not been realised. Additional research is needed to investigate the value of dMRI in combination with routinely collected data such as EEG. Further, the feasibility of user-friendly tools for deploying dMRI pipelines must be assessed.

Several works employ functional MRI (fMRI) to represent functional connectivity [10–12] alongside structural connectivity from dMRI. However, given EEG is routinely collected in epilepsy clinics, it may be a more accessible and practical alternative to fMRI, which has inherent, poor temporal resolution relative to EEG. White matter connectivity and information flow between specific brain regions has been linked to scalp EEG characteristics in healthy populations [13,14]. Further, EEG has been used to produce an individualised connectivity fingerprint that is robust across recordings [15], rendering its utility as a patient-specific, analytical network measure that can address the heterogeneous nature of focal epilepsy.

Discerning the seizure onset pattern and epileptogenic zone has been shown to improve the prognosis of post-surgical outcomes [16], and EEG and dMRI can aid this goal. A study on the role of scalp EEG in predicting post-surgical seizure outcomes showed abnormal MRI was valuable in ambiguous cases containing bilateral interictal epileptiform discharges [17], suggesting MRI may enhance prediction of seizure freedom. In another study of seven patients being evaluated for epilepsy, lesional and non-lesional MRIs were combined with high and low frequency bands from high density EEG (HDEEG) [18]. The Authors showed that the absence of structural support was related to significantly reduced functional connectivity in high frequency bands. Moreover, high frequency oscillations observed on scalp EEG are increasingly recognised as a hallmark of lesional epilepsy [19]. These works highlight the advantages of combining dMRI with EEG to detect aberrations that may typically only be partly revealed by one modality.

The majority of works that blend multimodal information from dMRI and EEG focus on source localisation techniques [20–22], using a digitiser to map electrode coordinates to the scalp which can be time-consuming. Others produced an automated, individualised localisation tool to map electrodes from high density EEG (HDEEG) to the scalp only, without extending the mapping to the cortex [23]. Many prior works favoured the combination of stereo EEG with dMRI [6,9,24,25], or only explored normal (non-ictal) awake EEG data with dMRI [26].

Several methods for electrical source localisation, which utilise a range of forward and inverse solutions, have been proposed and evaluated [27–29]. Thus the current study is distinguished from those prior works for the following reasons. We aimed to understand whether a patient-specific, structure-function coupling pattern could be observed without requiring manual digitisation of electrodes or applying one of the several forward and inverse solutions. We sought to apply our existing model [30], which maps cortex regions to individual electrodes, to a larger cohort. We specifically examined the seizure onset period (regardless of wakefulness state). Lastly, we aimed to validate the feasibility of our model as a clinically translatable method to leverage the potential of dMRI, with the view of elevating it to the established state currently held by structural MRI (i.e., T1) [31]. The dMRI component of our tool was designed to be deployed on a clinician's computer, allowing straightforward data processing from new patients (with ethics approval).

The contribution of this work is twofold: a. We extend the application of our spatial mapping model to a new patient cohort, highlighting consistent between-patient variance in region to electrode mapping, and b. We add to the growing body of research showing that connectivity data derived from structural MRI may augment scalp EEG observations for certain patients; acting as an additional tool during the diagnosis stage.

2. Materials and Methods

2.1. Participants and Data

Nine adults with focal epilepsy were recruited from the Comprehensive Epilepsy Centre at the Royal Prince Alfred Hospital (RPAH, Sydney, Australia), and MRI was performed at the Brain and Mind Centre (Sydney, Australia). Inclusion criteria were adults diagnosed with focal epilepsy, aged 18–60, presenting without surgery, with a minimum of two recorded seizures, and who were willing and able to comply with the study procedures for the duration of their participation. Exclusion criteria were pregnant women and individuals with intellectual disabilities. Ethical approval was obtained from the RPAH Local Health District (RPAH-LHD) ethics committee (see Institutional Review Statement in Section 5). The entire data processing and analysis consisted of several consecutive steps, depicted in Figure 1.

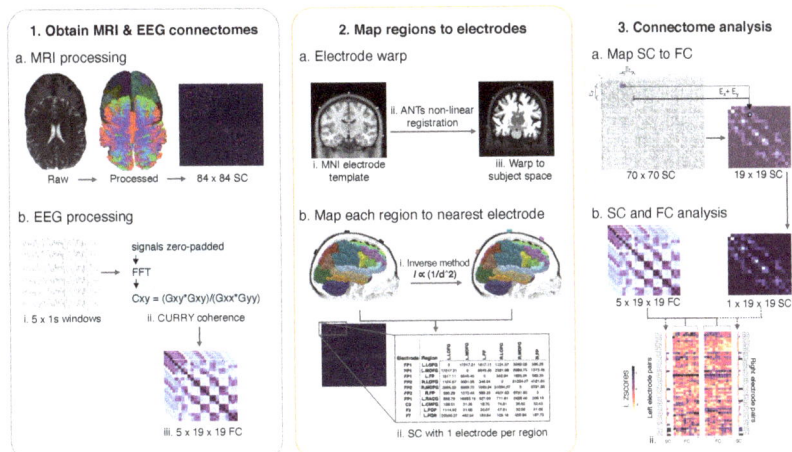

Figure 1. Schematic of data processing and analysis steps. To obtain the structural connectomes ("SC", Step 1, a), the dMRI was processed, and anatomically-constrained probabilistic tractography was conducted as outlined in Section 2.3. Cortical regions of interest were based on the Desikan-Killiany (DK) [32] atlas. To obtain the functional connectomes ("FC"), the first 5 s of a given seizure were selected using one-second windows. Each one-second window was processed using Curry's sensor coherence algorithm (Step 1, b, ii), producing a 21 × 21 coherence matrix. The reference electrodes were removed before statistical analysis, resulting in a 19 × 19 coherence matrix which was used as the functional connectome. In Step 2, the ANTs non-linear registration tool was used to warp electrodes in the MNI template space to the subject space, creating a subject-specific electrode warp (a, i–iii). To produce a subject-specific, one-to-one map of each cortical region to its nearest electrode, we applied our inverse square method (b, i). The inputs were each subject's electrode warp and cortical region labels from Step 1. The result was a structural connectome with an electrode name corresponding to each of the 70 regions, i.e., F7/L.LOFG (b, ii). In Step 3 (a), the 70 × 70 structural connectome was condensed to match the dimensions of the functional matrix (19 × 19). Specifically, the values of all regions corresponding to a given electrode pair were summed, and the total value was used as the connectivity value for that same electrode pair in the new condensed structural connectome. Lastly, z-scores were computed for all connectivity values in the structural and functional connectomes (b, i,ii) and the statistical analysis was conducted.

2.2. Image Acquisition

Image acquisition was described previously [33]. Briefly, all scans were acquired on the same GE Discovery™ MR750 3T scanner (GE Medical Systems, Milwaukee, WI, USA). The following sequences were acquired for each participant: Pre-contrast 3D high-resolution T1-weighted image (0.7 mm isotropic) using fast spoiled gradient echo (SPGR) with magnetisation-prepared inversion recovery pulse (TE/TI/TR=2.8/450/7.1 ms, flip angle = 12); and axial diffusion-weighted imaging (2 mm isotropic, TE/TR = 85/8325 ms) with a uniform gradient loading (b = 1000 s/mm^2) in 64 directions and 2 b0 s. An additional b0 image with reversed phase-encoding was also acquired for distortion correction [34].

2.3. Image Processing to Obtain Structural Connectomes

The T1 images were processed using a modified version of Freesurfer's recon-all (v6.0) [35], alongside an in-house skull-stripping tool (Sydney Neuroimaging Analysis Centre). Each subject was inspected, and minor segmentation errors were manually corrected. A 5 tissue-type (5TT) image [36] was generated using MRtrix3 [37]. The T1 image was registered to the mean b0 image; the warp was used to register the 5TT image, and the DK [32] parcellation image to the diffusion image.

Diffusion image processing was conducted using MRtrix3 [37]. The diffusion pre-processing included motion and distortion correction [34,38], bias correction using ANTs [39]. The dhollander algorithm [40] was used to estimate the response functions of the white matter, grey matter, and cerebral spinal fluid, from which constrained spherical deconvolution was used to estimate the fibre orientation distributions using MRtrix3Tissue [37]. The intensity of the white matter fibre orientation distributions was normalised [37], and used for anatomically constrained whole-brain tractography [41] (along with the registered 5TT image). The tractography specifications were as follows: 15 million tracks were generated, iFOD2 probabilistic fibre tracking [42], dynamic seeding [43], maximum length 300 mm, backtrack selected and crop at grey-matter-white-matter interface selected. For quantitative analysis, the corresponding weight for each streamline in the tractogram was derived using SIFT2 [43]. The streamlines and corresponding SIFT2 weights were used to create a weighted, undirected structural connectome ("SC") using the registered parcellation image. All image processing steps are shown in Figure 1 (Step 1, a).

2.4. EEG Acquisition

The EEG recordings were derived from ward recordings conducted during the patients' stay at the RPAH. The EEG was recorded using Compumedics hardware and software. The ward nurse applied the individual electrodes to the patient's head in the standard 10/20 format using the gold standard measurement process. Once the routine clinical recording was complete, the raw EEG files were obtained, and the seizure segments annotated by the EEG technician and reviewed by a senior neurologist. All seizures were then analysed in Curry 8 ("Curry", Compumedics Neuroscan) to obtain the functional connectomes. Curry is a neuroimaging software suite that allows the combination and analysis of multimodal data and is optimised for evaluating epilepsy-related data.

2.5. EEG Processing to Obtain Functional Connectomes

Curry was used to pre-process the EEG and obtain the sensor-based coherence matrices which represented the functional connectomes. First, we applied Curry's automated artifact reduction and filtering tool to obtain a clean signal. Next, Curry's coherence calculation process (shown in Figure 1, Step 1, b, ii) was used to generate coherence matrices for the first five seconds of each seizure. Specifically, using one-second non-overlapping windows starting from the annotated seizure onset time, the coherence matrices were computed from the cross-spectral densities Gxy and auto-spectral densities Gxx and Gyy of the channels x and y, using the equation

$Cxy = (Gxy \times Gxy)/(Gxx \times Gyy)$. The resulting coherence matrices were 21 × 21; row and column headers represented single electrodes. The reference electrodes and their corresponding scores were then removed, resulting in 19 × 19 matrices, which were used as the functional connectomes ("FC", Figure 1, Step 1, b, iii). Therefore, each electrode pair's corresponding value was a composite of the normalised maximum similarity between the waveforms and the time-shift (delay) when the maximum similarity occurred. The electrode pair value represented the highest percentage of coherence achieved by that electrode pair in the one-second window after factoring in the signal time lag between the two electrodes.

2.6. Mapping Cortical Regions to the Nearest Electrode

This section details the processes in Step 2 of Figure 1, where we used our previously described method [30] to create a subject-specific electrode warp and map each subject's cortical regions from the DK atlas to the nearest electrode. First, using the ANTs nonlinear registration tool, 21 electrodes in the standard MNI template space were warped to each participant's T1 image that had been registered to the diffusion image space (Figure 1 Step 2, a, i–iii). Next, we applied our inverse square method, which incorporates the inverse square equation shown in Figure 1 (Step 2, b, i) to produce a subject-specific, one-to-one mapping of each cortical region to its nearest electrode. The inputs were each subject's electrode warp and the cortical structure labels from Step 1. The inverse square equation holds that the light intensity of a source is inversely proportional to the square of the distance from the source. Thus, the inverse square method enabled the consideration of MRI voxel intensity in assessing the distance of each cortical region from each electrode's centre. Voxel intensity may represent the cortex's topological arrangement, endorsing postulation of the EEG signal strength from a given region relative to that region's distance from the scalp. The matrix in Figure 1, (Step 2, b, ii) depicts each region with one electrode name assigned—this electrode was the closest to that region. Subcortical regions (such as the hippocampus) were not assigned electrodes as their physical distance from the scalp and positioning below other cortical regions deemed them inaccessible for accurate measurement; thus they were removed from the analysis.

2.7. Mapping the Structural Connectome to the Functional Connectome

To enable the direct, one-to-one comparison of the values in the structural and functional connectomes, the structural connectome was first condensed to match the size of the functional connectomes (from 70 × 70 to 19 × 19). Only the upper triangle of the structural connectome was used in the calculation. The output file shown in Figure 1 (Step 2, b, ii) provided the electrode names and corresponding regions (and their values) for the new structural matrix. To calculate the new value for a given electrode pair in the condensed structural connectome, the values for all regions between that given electrode pair were summed. An example is provided in Figure 1, (Step 3, a), where all values between electrodes Fz ("E_x") and Fp2 ("E_y") are coloured in purple. The total sum of all values between E_x and E_y was used as the new value for Fz-Fp2 (black square) in the condensed structural connectome. Once new values were computed for all electrode pairs, the diagonal line (self correlations) was removed from the structural and functional connectomes, and the connectomes were converted to a 1D array for statistical analysis.

2.8. Statistical Analysis of Structure-Function Coupling

To test whether the laterality of the strong connections matched each patient's diagnosis, we first split the structural and functional connectomes into left and right hemispheres (Figure 1, Step 3, b) and removed cross-hemisphere electrode pairs. For example, if an electrode pair contained two electrodes in the left hemisphere (i.e., FP1-F3) or one left hemisphere and one central electrode (FP1-Fz), the electrode pair was kept. All electrode pairs that crossed from one hemisphere to the other (i.e., F3-F4) were removed. To test whether the highly connected electrode pairs from the structural and functional connectomes were congruent, we first computed the z-scores for all electrode pairs from all connectomes (Figure 1, Step 3, b, i). The z-score arrays were: a. 1 × 1D array per hemisphere for the structural connectome and b. 5 × 1D arrays (for each 1 s time window) per seizure, per hemisphere for the functional connectomes. Lastly, the structural and functional zscores from each hemisphere were displayed in parallel format for analysis (Figure 1, Step 3, b, ii).

To preserve only the most robust connections to represent high coherence between two electrodes, a z-score threshold of 2 (i.e., two standard deviations from the mean) was chosen for the structural connectome, and a threshold of 1.8 was chosen for the functional connectomes. Next, the z-scores from both connectome types were compared per one-second window from each seizure. If the same electrode pair from the structural and functional connectomes contained a z-score between 1.8 and 2 (or greater than 2), that electrode pair was classified as showing high structure-function coupling (termed "coupled electrode pairs").

The senior neurologist provided a "laterality" score for each patient based on whether the most frequently observed seizure onset zone was consistently restricted to one hemisphere. A laterality score of zero represented poorly lateralised seizures, whilst highly lateralised seizures received a score of three. If overall, the patient had late-lateralising seizures, they were classified as being non-lateralised at onset (i.e., a score of 0–1). Each patient was also assigned an "expected onset zone", predicated on the most frequently observed onset zone observed in all of a patient's recorded seizures, including seizures that were poorly or non-lateralised. The neurologist reviewed the raw EEG to confirm whether the electrode pair with high structure-function coupling was congruent with the expected seizure onset zone. All z-scores and statistical analyses were produced in SPSS v28 (Armonk, NY, USA: IBM Corp).

3. Results

3.1. Demographics

Nine patients (6F, mean age 38.8 ± 11.28) were included in this study after meeting the inclusion criteria. The patient characteristics, including seizure onset zone, are shown in Table 1. Three of the nine patients presented with highly lateralised seizures; the other six had a mixture of highly lateralised and poorly lateralised seizures. All patients were diagnosed with focal epilepsy; two had experienced frequent focal to bilateral tonic-clonic (FBTC) seizures, whilst another three had infrequent FBTC seizures (experienced more than one year prior to the EEG recording).

Table 1. Characteristics of patients.

Patient	Sex	Classification	MRI Diagnosis	Onset Age	Age at MRI	Duration	Drug Res.	Handedness
1	F	Left fronto-temporal	Normal	49	53	4	Y	R
2	M	Left fronto-temporal	HS [†]	21	49	28	Y	L
3	F	Right frontal	Normal	38	48	10	N	R
4	F	Right temporal	Normal	16	29	13	Y	U
5	M	Left fronto-temporal	Normal	16	31	15	Y	R
6	F	Left occipital	Normal	12	47	35	Y	R
7	F	Left fronto-temporal	Normal	35	48	13	N	R
8	M	Right fronto-temporal	Normal	15	33	18	Y	R
9	M	Right temporal	Normal [‡]	22	29	7	Y	R

Key: L: left, R: right, U: unknown; [†] HS: hippocampal sclerosis; [‡] slight enlargement of right amygdala.

3.2. Electrode-Region Mapping

The regions that displayed the most variance in electrode mapping across all nine patients are listed in Table 2 according to the Freesurfer region names. The majority of the variance appeared to be in the temporal regions. Manual inspection of the warped electrodes on each patient's scalp, which were overlaid on the cortex regions, indicated that individual scalp and cortex morphology contributed to the model's determination of the nearest electrode for a given region.

Table 2. Between patient region variance in the region to electrode mapping.

Subject No. Region Name	1	2	3	4	5	6	7	8	9
L. rostralanteriorcingulate	FP1	FP1	FP1	FZ	F3	FP1	FZ	FP1	FP1
R. rostralanteriorcingulate	FP2	FP2	FP2	FZ	FP2	FP2	FZ	F4	FP2
L. parsopercularis	F3	F7	F7	F3	F7	F7	T3	F7	F3
R. parsopercularis	F8	F8	F8	F8	F8	F8	F4	T4	F4
L. insula	T3	F7	T3	F7	T3	T3	T3	T3	T3
R. insula	T4	F8	F8	F8	F8	T4	C4	T4	T4
L. inferiortemporal	T5	T5	T5	T5	T3	T5	T5	T3	T5
R. inferiortemporal	T6	T4	T6	T6	T4	T6	T6	T4	T6
L. lateralorbitofrontal	FP1	F7	F7	F7	F7	F7	F7	F7	F7
R. lateralorbitofrontal	FP2	F8	F8	F8	F8	F8	F4	F8	FP2
L. cuneus	O1	O1	O1	O1	O1	O1	PZ	O1	O1
R. cuneus	O2	O2	O2	PZ	O2	O2	PZ	O2	O2
L. transversetemporal	T3	T3	T3	T3	T3	T3	T3	T3	T3
R. transversetemporal	T4	T4	T4	C4	T4	T4	T4	T4	T4
L. caudalanteriorcingulate	FZ	FZ	FZ	FZ	F3	FZ	FZ	F3	FZ
R. caudalanteriorcingulate	FZ	FZ	FZ	FZ	FZ	FZ	FZ	F4	FZ
L. isthmuscingulate	PZ	PZ	PZ	CZ	PZ	PZ	PZ	PZ	PZ
R. isthmuscingulate	PZ	PZ	PZ	CZ	PZ	PZ	PZ	PZ	PZ
R. bankssts	T6	T6	T4	T6	T4	T6	T6	T6	T6
R. superiorfrontal	FZ	FZ	FZ	FZ	FZ	FZ	FZ	CZ	FZ
R. caudalmiddlefrontal	C4	C4	C4	F4	C4	C4	C4	C4	C4
R. temporalpole	F8	F8	F8	F8	F8	F8	F8	T4	F8
L. supramarginal	C3	P3	C3	C3	C3	C3	C3	C3	C3
L. superiorparietal	P3	PZ	PZ	PZ	PZ	PZ	PZ	PZ	PZ

Key: L: left, R: right, bankssts: banks of the superior temporal sulcus.

3.3. Structure-Function Coupling

The structure-function coupling observed in the nine patients revealed three distinct groups with the following features. The first group (Patients 1–3, Figure 2a) had the highest laterality scores (L = 3), with coupled electrode pairs that consistently overlapped with the seizure onset zone. The second group (Patients 4–6, Figure 2b) had less well-lateralised seizures (L = 2–3), and the coupled electrode pairs overlapped with the onset side but not the exact zone. In Patient 4, only two out of three seizures were highly lateralised (L = 3) while the third was not (L = 2), and the electrode pair (PZ-O2) that did not overlap with the exact seizure onset zone was observed on the poorly lateralised seizure. Patient 4 also had three single electrodes from highly connected MRI and EEG pairs that overlapped inside the seizure onset zone (shown in heatmaps in Figure A1). Patients 5 and 6 were less well lateralised, and highly coupled electrode pairs were present both within and outside the seizure onset zone. In Patient 6, the electrode pair T3-C3 overlapped with the expected onset zone of the one poorly lateralised seizure. The third group (Patients 7–9, Figure 2c) were considered non-lateralised for most of their seizures. These patients generally displayed highly coupled electrode pairs that were inconsistent with the expected onset zone or had only a single overlapping MRI and EEG electrode rather than a pair (Patient 7). Figure 2 contains a condensed interpretation of the detailed results for each patient, shown in Appendix A. The overlapping electrode pairs with MRI z-scores (>2) and EEG z-scores (>1.8) are displayed. The most common seizure onset zone for each patient is also displayed.

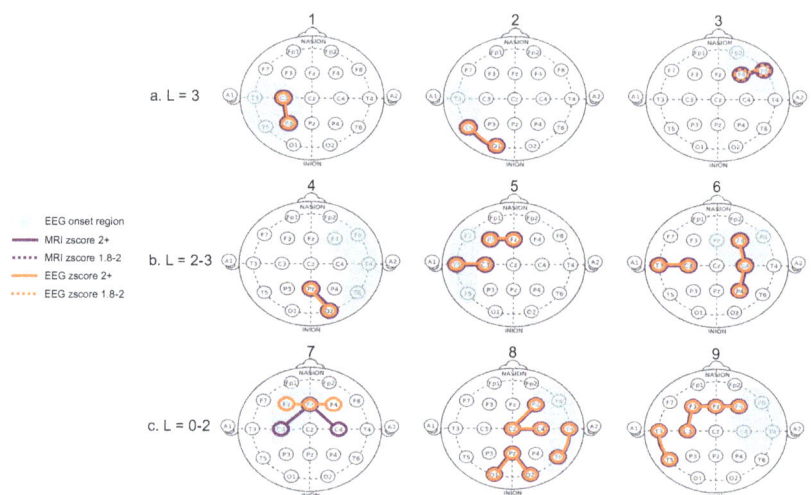

Figure 2. Highly connected electrode pairs in structural and functional connectome. Each "head" shows a schematic of the electrode pairs for each of the nine patients (numbered 1–9 above each head). The "L" value represents each patient's overall seizure lateralisation score based on their available recorded seizures. A score of zero represented poorly lateralised seizures, whilst highly lateralised seizures received a score of three. The patients' seizures stratified them into three categories: patients 1–3 had high laterality, patients 4–6 had some seizures that were well lateralised, while others were not, and patients 7–9 had poor laterality in all seizures. The purple lines (and circled electrodes) represent the electrode pairs that displayed strong connectivity (z-scores > 2) in the structural (MRI) connectome. The orange lines (and circled electrodes) represent the electrode pairs that displayed strong connectivity (z-scores > 2) in the functional (EEG) connectome. Dotted lines in either colour represent a z-score of 1.8–2. The blue shading represents the most frequently observed seizure onset zone for a given patient, as observed from their ward EEG recordings. If a seizure did not have a specific onset region within the first 5 s, it was considered non-lateralised, even if it displayed late-lateralisation. Purple and orange circled electrodes in the blue shaded areas represent high structure-function coupling in the seizure onset zone.

4. Discussion

In this study, we obtained structural and functional connectomes from nine patients to investigate whether our model could uncover the structure-function coupling during seizure onset. We also examined the pattern and congruence of the structure-function coupling with the expected seizure onset zone. The first key finding was that patients with well-lateralised seizures displayed high structural-functional congruence consistent with the expected seizure onset zone. The second key finding was that patients who were not well-lateralised had varying coupled electrodes that were not consistently in the onset zone. The results indicate that for well-lateralised patients, connectivity data derived from dMRI can be a valuable tool to augment routine EEG observations. However, the dMRI should be interpreted in the context of other routinely collected data from the patient.

Our findings offer some compelling evidence for the use of dMRI in clinical practice. Firstly, in patients with high structure-function coupling in the expected onset zone, dMRI may provide additional support to the EEG observations. A recent work used intracranial EEG (iEEG) and dMRI to explore the relation between structure-function coupling and post-surgery seizure freedom [9]. The Authors showed that patients who achieved post-surgery seizure freedom had higher structure-function coupling pre-surgery. However, access to iEEG may not be feasible in the initial diagnosis stage. Additionally, the diagnostic yield of low-density scalp EEG (25 electrodes) has been suggested to be comparable to high-density EEG (256 electrodes) [44]. Taken together, our findings suggest that our model may be used during the diagnosis stage to determine the suitability of surgery for newly-diagnosed patients. Notably, we provide evidence that high structure-function coupling was present for patients regardless of whether they had previously experienced an FBTC seizure. Our finding suggests that a history of infrequent FBTC seizures (present in Patients 1 and 3) does not preclude the patient from having well-lateralised structure-function coupling.

Secondly, in the patients with poorly lateralised seizures, the structure-function coupling was more predominant in the ipsilateral hemisphere. However, the presence of structure-function coupling in the contralateral hemisphere was unsurprising, given their laterality score. In these patients, dMRI may provide additional information that can guide the placement of additional electrodes in longer ward recordings. However, a more extensive structure-function coupling model may be needed to understand whether the poor laterality can be attributed to the equally high structural connectivity in both hemispheres or some other biophysical phenomenon. Further, our model demonstrated a specific region with high structure-function coupling for Patient 8, who was initially considered poorly lateralised on EEG. Such cases highlight that dMRI may offer endorsement of a specific onset zone to support an otherwise inconclusive EEG recording.

Interestingly, the presence of high structure-function coupling in an electrode pair containing a middle electrode (FZ, CZ, PZ) was observed in several patients with a seizure laterality score of 2 or lower. For example, patients 5, 7, 8 and 9 had at least one middle electrode in the electrode pairs that showed high structure-function coupling. Their poor seizure lateralisation could be due to the electrophysiological activity beneath the middle electrode, which may drive the contralateral seizure propagation.

Given the scope of the current work, some methodological considerations may aid the interpretation of the results. Firstly, the mapping method did not appear to impact the results as well-lateralised patients had a similar number of variances in the electrode-to-region mapping as the less well-lateralised patients, for whom the structural data provided little additional information. However, laterality alone may not account for the results from patients who displayed high structure-function coupling congruent with the expected onset zone. Our selection of time window and bandwidth may have impacted the coherence score. The one-second window was perhaps not brief enough to capture the highly coherent initial EEG activity. We observed high coherence in the ipsilateral hemisphere at the 0–100 millisecond (ms) scale for some patients. The lengthy time window may have confounded this high coherence.

Further, the proposition that microscopic signal aberrations in EEG may not be observed from a macroscopically normal EEG signal is worth considering [2]. The uncertainty of empirical, visual evaluation of EEG to localise the seizure onset zone has been shown [45]. Thus, in the current work, the true onset zone for some individuals may not be observable on the raw EEG, yet may have been captured in the processed EEG coherence data. However, such postulations must be verified using high-definition EEG or intracranial/stereo EEG. Alternatively, since we combined microscopic MRI and EEG data, it is also possible that our model captured the genuine seizure onset zone. Supposing the seizure began in a different region, it could have fused in millisecond time with other active regions and thus was visually revealed in a different region on the raw EEG. Such an explanation is conceivable for individuals with FBTC or multi-onset seizures and is a topic for future investigation.

Lastly, it is feasible that the structural topology has a diminished relationship with the functional activity in some individuals. Numerous mechanisms mould the seizure propagation pathways, and within-patient variance has been shown [46]. Lateralisation may be inextricably linked to the seizure duration. However, based on our experimental design (i.e., time window selection) and modest sample size, it was not plausible to extrapolate any biophysical mechanisms that may be in force. The individuals in the third group may have less well-defined circuits, or a different epileptic pattern, i.e., multi-onset or deep structural connections or functional activity, that was not captured in the connectome reconstruction or on the scalp EEG. These concepts are the subject of ongoing work to further elaborate on this feasibility study and better explain the individual differences in the outcomes.

Although our methodology may explain some of the findings, potential limitations were inevitable. The generalisation of the results is limited by the small sample size, endorsing the need to extend this work to a larger cohort. The placement of the electrodes in the electrode warp and automated mapping cannot be deemed identical to the original, physical placement of the electrodes during the EEG recording. The electrode warp placements represented the expected actual electrode placement during the recording. This work highlights the possible variation inherent in demystifying the electrical signals captured on scalp EEG. There is an intrinsic between-patient variance in cortex and scalp morphology and thickness. The nature of the clinical procedure introduces further variance through the measurement estimation of different technicians during the application and re-application of electrodes.

The Curry sensor coherence algorithm is confined to broadband frequencies and does not compute coherence from narrow-band frequencies. Further, the Curry algorithm does not consider the spatial, topographical, morphological or biophysical implications of the scalp signal; it is calculated purely on the raw EEG wave. Therefore, our inverse square mapping method was constrained in accounting for the spatial and biophysical properties of the scalp signal measurements. It is possible that the single neurologist's assessment of the congruency results may introduce bias. Future work will include evaluation of the EEG data by several neurologists blinded to the methodology, and an inter-rater reliability assessment (such as the Kappa statistic). Lastly, lack of control data restricts the distinction between the structure-function coupling in seizures and normal coupling in resting state or non-ictal periods.

Indeed, linking brain structure and function remains an imperfect science, confounded by individual differences in structure-function coupling [47]. Therefore extending this study to a larger cohort, with the addition of control data, is the subject of ongoing work in our lab. The inclusion of pre-ictal and cross-hemisphere connectivity data will enable further comparison and quantification of the active electrodes across varying brain states. Additionally, using coherence matrices derived from open source software could provide a point of comparison to Curry's sensor coherence maps. Despite these restrictions, our work provides evidence that dMRI is a promising additional tool to classify patients for further investigation or surgical candidature. Our model may be practical in identifying the most active locations for sub-scalp electrodes and the patients who could benefit from ultra long-term monitoring. We show that all regions of high connectivity are not necessarily the best place for sub-scalp electrodes. We present a feasible method to distinguish patients, and patient-specific brain regions, that may be candidates for sub-scalp electrodes. With further refinement, our method could be utilised in identifying the optimal position for sub-scalp electrode placement, removing the need for more invasive EEG methods.

5. Conclusions

In conclusion, this study utilised a model to spatially map scalp electrodes to the nearest brain region and compare the structural and functional connectivity in nine patients with focal epilepsy. We showed that not all highly connected structural regions result in highly connected scalp EEG in the same region. Our findings suggest that seizures may follow strong connections intermittently and might only do so in well-lateralised patients and not for every seizure. Less well-lateralised patients displayed some high structure-function coupling in the ipsilateral hemisphere, but this was inconsistent. Our findings contribute to the evidence supporting the use of dMRI in clinical practice, which can guide patient-specific electrode placement and enhance the detection of the seizure onset zone. Future work will include comparisons with open source software and the addition of interictal and control data.

Author Contributions: Conceptualisation, C.M., C.W. and A.N.; methodology, C.M., C.W. and A.N.; clinical data acquisition, C.M. and A.N.; data curation, C.M. and A.D.; diffusion imaging pipeline, implementation and analysis, C.M., A.D. and C.W.; statistical analysis, C.M. and C.W.; manuscript writing, C.M.; manuscript revision and editing C.M., A.D., M.B., O.K., C.W. and A.N.; clinical advisory and results interpretation, A.N. All authors have read and agreed to the published version of the manuscript.

Funding: This research received no specific external funding.

Institutional Review Board Statement: All research and methods were performed in accordance with the Declaration of Helsinki, and the relevant guidelines and regulations prescribed by the RPAH-LHD and ethics committees. The study was approved by the Ethics Committee from the RPAH Local Health District (RPAH-LHD). The protocol number and ethics approval ID for the MRI data are X14-0347 and HREC/14/RPAH/467. The protocol number and ethics approval ID for the EEG data are X19-0323 and 2019/ETH11868.

Informed Consent Statement: The requirement for informed consent was waived in the approved ethics for the EEG data (protocol number and approval ID are X19-0323 and 2019/ETH11868) since only de-identified EEG data was acquired. Written informed consent was obtained from all participants who attended the Brain and Mind Centre for an MRI scan, as per the approved MRI ethics (protocol number and approval ID are X14-0347 and HREC/14/RPAH/467).

Data Availability Statement: The datasets generated and/or analysed during the current study are not publicly available because they are RPAH patients and can only be accessed by authorised individuals named on the approved ethics. However, de-identified, processed data can be made available upon request to the corresponding author, and subject to approval from the governing ethics entities at the RPAH and The University of Sydney.

Acknowledgments: The Authors acknowledge all staff at the Comprehensive Epilepsy Centre at the RPAH, particularly Maricar Senturias (RN/ACNC Epilepsy), who assisted with patient recruitment. The Authors acknowledge the radiology staff at i-MED Radiology for their assistance in obtaining the MRI data. The Authors acknowledge the research funding support from UCB Australia Pty Ltd. C.M. acknowledges scholarship support from the Nerve Research Foundation, University of Sydney. A.D. acknowledges funding from St. Vincent's Hospital. O.K. acknowledges the partial support provided by The University of Sydney through a SOAR Fellowship and Microsoft's partial support through a Microsoft AI for Accessibility grant. CW acknowledges research funding from the Nerve Research Foundation, University of Sydney.

Conflicts of Interest: The Authors declare no conflict of interest. The funders had no role in the design of the study; in the collection, analyses, or interpretation of data; in the writing of the manuscript; or in the decision to publish the results.

Appendix A

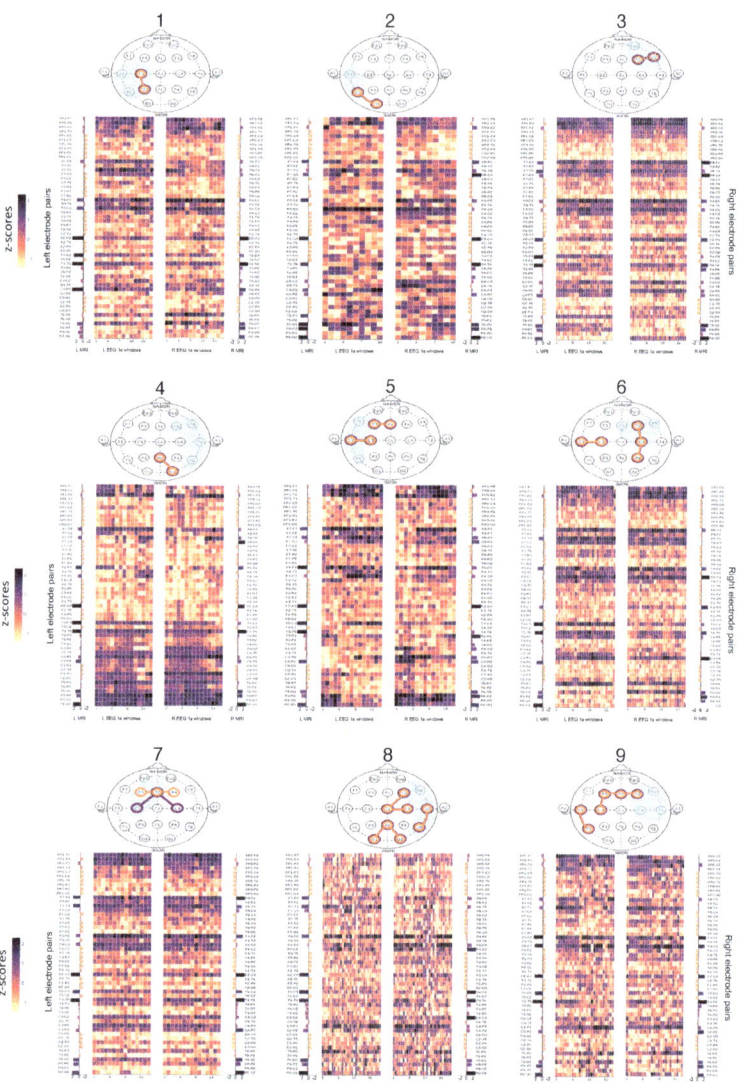

Figure A1. Heatmaps of z-scores from MRI and EEG connectomes. L: Left, R: Right. Patients are numbered 1–9 above each "head" and their respective connectomes are shown below the "head" map in the same grouping as Figure 2 in the main text. The "EEG 1s windows" depict the left and right side of each EEG connectome, split into 1 s windows. The structural (MRI) connectomes are split into left and right sides ("L MRI" and "R MRI") and positioned alongside the matching EEG connectome side (i.e., left MRI next to left EEG connectome). The left and right electrode pairs are listed next to the respective side of the connectome.

References

1. Weng, Y.; Larivière, S.; Caciagli, L.; Vos de Wael, R.; Rodríguez-Cruces, R.; Royer, J.; Xu, Q.; Bernasconi, N.; Bernasconi, A.; Thomas Yeo, B.; et al. Macroscale and microcircuit dissociation of focal and generalized human epilepsies. *Commun. Biol.* **2020**, *3*, 1–11. [CrossRef] [PubMed]
2. Zijlmans, M.; Zweiphenning, W.; van Klink, N. Changing concepts in presurgical assessment for epilepsy surgery. *Nat. Rev. Neurol.* **2019**, *15*, 594–606. [CrossRef] [PubMed]
3. Babaeeghazvini, P.; Rueda-Delgado, L.M.; Gooijers, J.; Swinnen, S.P.; Daffertshofer, A. Brain Structural and Functional Connectivity: A Review of Combined Works of Diffusion Magnetic Resonance Imaging and Electro-Encephalography. *Front. Hum. Neurosci.* **2021**, *15*, 585. [CrossRef] [PubMed]
4. Friston, K.J. Functional and effective connectivity in neuroimaging: A synthesis. *Hum. Brain Mapp.* **1994**, *2*, 56–78. [CrossRef]
5. Honey, C.J.; Sporns, O.; Cammoun, L.; Gigandet, X.; Thiran, J.P.; Meuli, R.; Hagmann, P. Predicting human resting-state functional connectivity from structural connectivity. *Proc. Natl. Acad. Sci. USA* **2009**, *106*, 2035–2040. [CrossRef] [PubMed]
6. Shah, P.; Ashourvan, A.; Mikhail, F.; Pines, A.; Kini, L.; Oechsel, K.; Das, S.R.; Stein, J.M.; Shinohara, R.T.; Bassett, D.S.; et al. Characterizing the role of the structural connectome in seizure dynamics. *Brain* **2019**, *142*, 1955–1972. [CrossRef]
7. Vattikonda, A.N.; Hashemi, M.; Sip, V.; Woodman, M.M.; Bartolomei, F.; Jirsa, V.K. Identifying spatio-temporal seizure propagation patterns in epilepsy using Bayesian inference. *Commun. Biol.* **2021**, *4*, 1–13. [CrossRef] [PubMed]
8. Lascano, A.M.; Perneger, T.; Vulliemoz, S.; Spinelli, L.; Garibotto, V.; Korff, C.M.; Vargas, M.I.; Michel, C.M.; Seeck, M. Yield of MRI, high-density electric source imaging (HD-ESI), SPECT and PET in epilepsy surgery candidates. *Clin. Neurophysiol.* **2016**, *127*, 150–155. [CrossRef] [PubMed]
9. Sinha, N.; Duncan, J.S.; Diehl, B.; Chowdhury, F.A.; de Tisi, J.; Miserocchi, A.; McEvoy, A.W.; Davis, K.A.; Vos, S.B.; Winston, G.P.; et al. Intracranial EEG structure-function coupling predicts surgical outcomes in focal epilepsy. *arXiv* **2022**, arXiv:2204.08086. [CrossRef]
10. Chiang, S.; Stern, J.M.; Engel, J., Jr.; Haneef, Z. Structural–functional coupling changes in temporal lobe epilepsy. *Brain Res.* **2015**, *1616*, 45–57. [CrossRef]
11. O'Muircheartaigh, J.; Keller, S.S.; Barker, G.J.; Richardson, M.P. White matter connectivity of the thalamus delineates the functional architecture of competing thalamocortical systems. *Cereb. Cortex* **2015**, *25*, 4477–4489. [CrossRef] [PubMed]
12. Morgan, V.L.; Sainburg, L.E.; Johnson, G.W.; Janson, A.; Levine, K.K.; Rogers, B.P.; Chang, C.; Englot, D.J. Presurgical temporal lobe epilepsy connectome fingerprint for seizure outcome prediction. *Brain Commun.* **2022**, *4*, fcac128. [CrossRef] [PubMed]
13. Gong, J.; Luo, C.; Chang, X.; Zhang, R.; Klugah-Brown, B.; Guo, L.; Xu, P.; Yao, D. White matter connectivity pattern associate with characteristics of scalp EEG signals. *Brain Topogr.* **2017**, *30*, 797–809. [CrossRef] [PubMed]
14. Deslauriers-Gauthier, S.; Lina, J.M.; Butler, R.; Whittingstall, K.; Gilbert, G.; Bernier, P.M.; Deriche, R.; Descoteaux, M. White matter information flow mapping from diffusion MRI and EEG. *NeuroImage* **2019**, *201*, 116017. [CrossRef]
15. Sadaghiani, S.; Brookes, M.J.; Baillet, S. Connectomics of human electrophysiology. *NeuroImage* **2022**, *247*, 118788. [CrossRef] [PubMed]
16. Lagarde, S.; Buzori, S.; Trebuchon, A.; Carron, R.; Scavarda, D.; Milh, M.; McGonigal, A.; Bartolomei, F. The repertoire of seizure onset patterns in human focal epilepsies: Determinants and prognostic values. *Epilepsia* **2019**, *60*, 85–95. [CrossRef] [PubMed]
17. Fitzgerald, Z.; Morita-Sherman, M.; Hogue, O.; Joseph, B.; Alvim, M.K.; Yasuda, C.L.; Vegh, D.; Nair, D.; Burgess, R.; Bingaman, W.; et al. Improving the prediction of epilepsy surgery outcomes using basic scalp EEG findings. *Epilepsia* **2021**, *62*, 2439–2450. [CrossRef] [PubMed]
18. Chu, C.J.; Tanaka, N.; Diaz, J.; Edlow, B.L.; Wu, O.; Hämäläinen, M.; Stufflebeam, S.; Cash, S.S.; Kramer, M.A. EEG functional connectivity is partially predicted by underlying white matter connectivity. *Neuroimage* **2015**, *108*, 23–33. [CrossRef] [PubMed]
19. Cserpan, D.; Gennari, A.; Gaito, L.; Lo Biundo, S.P.; Tuura, R.; Sarnthein, J.; Ramantani, G. Scalp HFO rates are higher for larger lesions. *Epilepsia Open* **2022**, *7*, 496–503. [CrossRef]
20. Samadzadehaghdam, N.; Makkiabadi, B.; Masjoodi, S.; Mohammadi, M.; Mohagheghian, F. A new linearly constrained minimum variance beamformer for reconstructing EEG sparse sources. *Int. J. Imaging Syst. Technol.* **2019**, *29*, 686–700. [CrossRef]
21. Rubega, M.; Carboni, M.; Seeber, M.; Pascucci, D.; Tourbier, S.; Toscano, G.; Van Mierlo, P.; Hagmann, P.; Plomp, G.; Vulliemoz, S.; et al. Estimating EEG source dipole orientation based on singular-value decomposition for connectivity analysis. *Brain Topogr.* **2019**, *32*, 704–719. [CrossRef] [PubMed]
22. Neugebauer, F.; Antonakakis, M.; Unnwongse, K.; Parpaley, Y.; Wellmer, J.; Rampp, S.; Wolters, C.H. Validating EEG, MEG and combined MEG and EEG beamforming for an estimation of the epileptogenic zone in focal cortical dysplasia. *Brain Sci.* **2022**, *12*, 114. [CrossRef] [PubMed]
23. Zagorchev, L.; Brueck, M.; Fläschner, N.; Wenzel, F.; Hyde, D.; Ewald, A.; Peters, J. Patient-specific sensor registration for electrical source imaging using a deformable head model. *IEEE Trans. Biomed. Eng.* **2020**, *68*, 267–275. [CrossRef] [PubMed]
24. Proix, T.; Bartolomei, F.; Guye, M.; Jirsa, V.K. Individual brain structure and modelling predict seizure propagation. *Brain* **2017**, *140*, 641–654. [CrossRef] [PubMed]
25. Jirsa, V.K.; Proix, T.; Perdikis, D.; Woodman, M.M.; Wang, H.; Gonzalez-Martinez, J.; Bernard, C.; Bénar, C.; Guye, M.; Chauvel, P.; et al. The virtual epileptic patient: Individualized whole-brain models of epilepsy spread. *Neuroimage* **2017**, *145*, 377–388. [CrossRef] [PubMed]

26. Varatharajah, Y.; Berry, B.; Joseph, B.; Balzekas, I.; Pal Attia, T.; Kremen, V.; Brinkmann, B.; Iyer, R.; Worrell, G. Characterizing the electrophysiological abnormalities in visually reviewed normal EEGs of drug-resistant focal epilepsy patients. *Brain Commun.* **2021**, *3*, fcab102. [CrossRef] [PubMed]
27. Asadzadeh, S.; Rezaii, T.Y.; Beheshti, S.; Delpak, A.; Meshgini, S. A systematic review of EEG source localization techniques and their applications on diagnosis of brain abnormalities. *J. Neurosci. Methods* **2020**, *339*, 108740. [CrossRef] [PubMed]
28. Sharma, P.; Seeck, M.; Beniczky, S. Accuracy of interictal and ictal electric and magnetic source imaging: A systematic review and meta-analysis. *Front. Neurol.* **2019**, *10*, 1250. [CrossRef]
29. Birot, G.; Spinelli, L.; Vulliémoz, S.; Mégevand, P.; Brunet, D.; Seeck, M.; Michel, C.M. Head model and electrical source imaging: A study of 38 epileptic patients. *NeuroImage Clin.* **2014**, *5*, 77–83. [CrossRef]
30. Maher, C.; D'Souza, A.; Zeng, R.; Wang, D.; Barnett, M.; Kavehei, O.; Armin, N.; Wang, C. Automated method to map cortical brain regions to the nearest scalp electroencephalography electrode. *Epilepsia* **2021** *62*, 238–239
31. Rados, M.; Mouthaan, B.; Barsi, P.; Carmichael, D.; Heckemann, R.A.; Kelemen, A.; Kobulashvili, T.; Kuchukhidze, G.; Marusic, P.; Minkin, K.; et al. Diagnostic value of MRI in the presurgical evaluation of patients with epilepsy: influence of field strength and sequence selection: a systematic review and meta-analysis from the E-PILEPSY Consortium. *Epileptic Disord.* **2022**, *24*, 323–342. [CrossRef] [PubMed]
32. Desikan, R.S.; Ségonne, F.; Fischl, B.; Quinn, B.T.; Dickerson, B.C.; Blacker, D.; Buckner, R.L.; Dale, A.M.; Maguire, R.P.; Hyman, B.T.; Albert, M.S. An automated labeling system for subdividing the human cerebral cortex on MRI scans into gyral based regions of interest. *NeuroImage* **2006**, *31*, 968–980. [CrossRef] [PubMed]
33. Maher, C.; D'Souza, A.; Zeng, R.; Barnett, M.; Kavehei, O.F.; Nikpour, A.; Wang, C. White matter alterations in focal to bilateral tonic-clonic seizures. *Front. Neurol.* **2022**, *13*, 972590. [CrossRef]
34. Andersson, J.L.; Sotiropoulos, S.N. An integrated approach to correction for off-resonance effects and subject movement in diffusion MR imaging. *NeuroImage* **2016**, *125*, 1063–1078. [CrossRef]
35. Fischl, B. Freesurfer. *Neuroimage* **2012**, *62*, 774–781. [CrossRef]
36. Smith, R.; Skoch, A.; Bajada, C.J.; Caspers, S.; Connelly, A. Hybrid surface-volume segmentation for improved anatomically-constrained tractography. In Proceedings of the OHBM Annual Meeting, Virtual, 23 June–3 July 2020; pp. 1–5.
37. Tournier, J.D.; Smith, R.; Raffelt, D.; Tabbara, R.; Dhollander, T.; Pietsch, M.; Christiaens, D.; Jeurissen, B.; Yeh, C.H.; Connelly, A. MRtrix3: A fast, flexible and open software framework for medical image processing and visualisation. *NeuroImage* **2019**, *202*, 116137. [CrossRef] [PubMed]
38. Smith, S.M.; Jenkinson, M.; Woolrich, M.W.; Beckmann, C.F.; Behrens, T.E.; Johansen-Berg, H.; Bannister, P.R.; De Luca, M.; Drobnjak, I.; Flitney, D.E.; et al. Advances in functional and structural MR image analysis and implementation as FSL. *NeuroImage* **2004**, *23*, S208–S219. [CrossRef] [PubMed]
39. Tustison, N.J.; Avants, B.B.; Cook, P.A.; Zheng, Y.; Egan, A.; Yushkevich, P.A.; Gee, J.C. N4ITK: Improved N3 bias correction. *IEEE Trans. Med. Imaging* **2010**, *29*, 1310–1320. [CrossRef]
40. Dhollander, T.; Raffelt, D.; Connelly, A. Unsupervised 3-tissue response function estimation from single-shell or multi-shell diffusion MR data without a co-registered T1 image. In Proceedings of the ISMRM Workshop on Breaking the Barriers of Diffusion MRI. ISMRM, Lisbon, Portugal, 12–16 September 2016; Volume 5.
41. Smith, R.E.; Tournier, J.D.; Calamante, F.; Connelly, A. Anatomically-constrained tractography: Improved diffusion MRI streamlines tractography through effective use of anatomical information. *NeuroImage* **2012**, *62*, 1924–1938. [CrossRef]
42. Tournier, J.D.; Calamante, F.; Connelly, A. Improved probabilistic streamlines tractography by 2nd order integration over fibre orientation distributions. In Proceedings of the International Society for Magnetic Resonance in Medicine, Stockholm, Sweden, 1–7 May 2010; p. 1670.
43. Smith, R.E.; Tournier, J.D.; Calamante, F.; Connelly, A. SIFT2: Enabling dense quantitative assessment of brain white matter connectivity using streamlines tractography. *NeuroImage* **2015**, *119*, 338–351. [CrossRef]
44. Justesen, A.B.; Foged, M.T.; Fabricius, M.; Skaarup, C.; Hamrouni, N.; Martens, T.; Paulson, O.B.; Pinborg, L.H.; Beniczky, S. Diagnostic yield of high-density versus low-density eeg: The effect of spatial sampling, timing and duration of recording. *Clin. Neurophysiol.* **2019**, *130*, 2060–2064. [CrossRef] [PubMed]
45. Davis, K.A.; Devries, S.P.; Krieger, A.; Mihaylova, T.; Minecan, D.; Litt, B.; Wagenaar, J.B.; Stacey, W.C. The effect of increased intracranial EEG sampling rates in clinical practice. *Clin. Neurophysiol.* **2018**, *129*, 360–367. [CrossRef] [PubMed]
46. Schroeder, G.M.; Chowdhury, F.A.; Cook, M.J.; Diehl, B.; Duncan, J.S.; Karoly, P.J.; Taylor, P.N.; Wang, Y. Multiple mechanisms shape the relationship between pathway and duration of focal seizures. *Brain Commun.* **2022**, *4*, fcac173. [CrossRef] [PubMed]
47. Suárez, L.E.; Markello, R.D.; Betzel, R.F.; Misic, B. Linking structure and function in macroscale brain networks. *Trends Cogn. Sci.* **2020**, *24*, 302–315. [CrossRef]

Article

Building Networks with a New Cross-Bubble Transition Entropy for Quantitative Assessment of Mental Arithmetic Electroencephalogram

Xiaobi Chen [1], Guanghua Xu [1,2,*], Sicong Zhang [1], Xun Zhang [1] and Zhicheng Teng [1]

1 School of Mechanical Engineering, Xi'an Jiaotong University, Xi'an 710049, China
2 State Key Laboratory for Manufacturing Systems Engineering, Xi'an Jiaotong University, Xi'an 710049, China
* Correspondence: ghxu@xjtu.edu.cn

Abstract: The complex network nature of human brains has led an increasing number of researchers to adopt a complex network to assess the cognitive load. The method of constructing complex networks has a direct impact on assessment results. During the process of using the cross-permutation entropy (CPE) method to construct complex networks for cognitive load assessment, it is found that the CPE method has the shortcomings of ignoring the transition relationship between symbols and the analysis results are vulnerable to parameter settings. In order to address this issue, a new method based on the CPE principle is proposed by combining the advantages of the transition networks and the bubble entropy. From an interaction perspective, this method suggested that the node-wise out-link transition entropy of the cross-transition network between two time series is used as the edge weight to build a complex network. The proposed method was tested on the unidirectional coupled Henon model and the results demonstrated its suitability for the analysis of short time series by decreasing the influence of the embedding dimension and improving the reliability under the weak coupling conditions. The proposed method was further tested on the publicly available EEG dataset and showed significant superiority compared with the conventional CPE method.

Keywords: cognitive load; coupling; bubble entropy; transition network

1. Introduction

Different levels of cognitive demand can accommodate the complexity and variability of the everyday tasks and the environments, and can result in different cognitive loads [1–3]. Continuous high cognitive load will not only lead to inefficient work but also accidents that might lead to life-threatening consequences. In addition, it also has negative effects on physical and mental health, such as insomnia, decreased immunity, susceptibility to infection, and migraines [4–8]. As a practical necessity, the evaluation of cognitive load or mental load has become a hot topic of research. Therefore, it is of practical significance to design and build a system capable of detecting cognitive load. The use of such a system will not only make it possible to assess the impact of different tasks on the cognitive load, but more importantly, a timely and accurate estimate of cognitive load will help to determine the optimum level of mental load, in order to prevent accidents and make workers more compatible with the work environment. Conventionally, the measurement of cognitive load can be divided into subjective and objective measures [9]. Subjective measures are collected via interviews or questionnaires. They are usually unreliable due to the subjective opinions of the participants [10–12]. In contrast, objective measures that are mainly based on task performances or derived from physiological recordings are less intrusive to the task and independent of the participants' opinion. With the development of technology, neurophysiological activities from brain, heart, and eye movement can be recorded and analyzed to reflect the mental state objectively in a noninvasive way [13]. Previous studies have confirmed that signals such as near-infrared spectroscopy (NIRS), functional magnetic resonance imaging (fMRI), electrocorticography (ECoG), or electroencephalography (EEG) are closely correlated with brain status and can provide a useful way to assess cognitive load [14–18]. Among these physiological signals, EEG has been widely concerned by researchers because of its high time resolution, noninvasiveness, convenience, security, cheapness, and portability [19,20].

In general, the EEG signal is nonstationary and nonlinear. Linear analysis techniques in the time–frequency domain can be used to detect rhythmic oscillations, but the contained nonlinear information cannot be effectively extracted [21]. Therefore, many scholars have attempted to extract various nonlinear parameters from EEG signals and combine them with the machine learning technique in order to effectively capture the subtle information related to the physiological states. Nilima Salankar et al. used the empirical mode decomposition (EMD) and the variational mode decomposition (VMD) to decompose the EEG signals, respectively, and then used the second-order difference plots for feature mining of the decomposed intrinsic modes. The results showed that alcoholic (A) and nonalcoholic (NA) subjects could be accurately classified when using short-duration EEG recordings [22]. Mohammad Shahbakhti et al. proposed extracting Katz and Higuchi's fractal dimensions, dispersion entropy, and bubble entropy from the sub-band of a single-channel frontal EEG recording to construct the nonlinear feature set and then differentiate between the arousal and the sleep stage I [23]. Jose Kunnel Paul et al. used seven nonlinear parameters, including the sample entropy (SampEn), fractal dimension (FD), higher-order spectrum (HOS), maximum Lyapunov index (LLE), Kolmogorov complexity (KC), Hurst index (HE), and the band power of the EEG signal in sleep stage 2 and 3 as the features to classify between patients with fibromyalgia and healthy controls. The accuracy, sensitivity, and specificity of the classification results were 96.15%, 96.88%, and 95.65%, respectively [24]. The nonlinear parameters in the above-mentioned methods were taken from individual EEG channels and involve no information on the interaction between different channels. However, previous research has shown that the brain should be treated as a complex network system based on the many features it shares with networks of other biological and physical systems [25]. Complex network analysis is a powerful technique based on the graph theory that typically uses a small number of valid and reliable measures to capture the features of the brain network [26]. There is a growing interest in the cognitive load assessment through the construction of complex networks, and various methods have been proposed to convert time series into networks [27–30]. Complex networks constructed using different network construction algorithms may have distinct, significantly different properties [31]. A variety of methods have been proposed so far to define the concept of connectivity between nonlinearly coupling components and investigate the characteristics of the topological properties of networks. Among different methods, for example, the mutual information (MI) (including its time-delayed version) [32,33], transfer entropy (TE) [34], inner composition alignment (IOTA) [35] and cross-sample entropy (CSE) [36], the TE is widely used in particular as a nonparametric measure that does not rely on any assumption of some model and can capture the directional and dynamic interaction between the different components of a time series [37,38]. However, in practice, an unavoidable pitfall of TE is that robust estimation of the interactions requires long-term data recordings. In order to meet the need for interaction estimation using finite data samples, Shi et al. proposed the CPE by fusing inner composition alignment (IOTA) and permutation entropy, and validated it in financial time series analysis [39], noting that CPE was simple, stable, and efficient.

In the original CPE method, only the probability distribution of the symbols after coarse graining of the affected time series is considered during the calculation of entropy, ignoring the transition relationship between the symbols in the temporal domain. For example, given the symbolized set A = [2 2 4 3 5 1 2] and B = [1 2 2 5 3 2 4], the probability distributions of the elements in set A and set B are the same and, therefore, the original CPE method would obtain the same entropy value. In addition, like other nonlinear measures, the CPE method involves the manual selection of parameters to ensure the effectiveness of the results. In order to address these issues, a new method to construct the complex network based on the cross-transition network was proposed in this study to assess cognitive load. The novelty of the method lies in incorporating the advantages of transition network and bubble entropy [40] into the CPE to estimate the coupling strength of two time series from a cross-network perspective. The node-wise out-link transition entropy of two time series cross-transition networks was proposed as the edge weights between two time series to construct the complex network, and the network parameters were extracted as a quantitative measurement of the cognitive load. Referring to the symbolization process of the bubble entropy, the number of swaps required to sort the phase space reconstruction vectors of the affected time series in the ascending order was used instead of the number of intersections calculated by the OITA method in the original CPE. In order to verify the effectiveness of the proposed method, the unidirectional coupled Honen model with different coupling strengths was used, and the results were compared with those obtained using the original CPE. The proposed method and the original CPE method were further compared by constructing the complex network on the realistic EEG recordings from the mental arithmetic task. The significance of the selected network indicator and the capability of the proposed method to differentiate different levels of brain cognitive load were verified using the nonparametric permutation test.

The contributions of this paper are as follows.

1. Based on the cross-transition network, a novel method is proposed that reflects the information interaction between two time series in more detail.

2. The symbolization process with reference to the bubble entropy minimizes the effect of parameter setting on the analysis results.

3. The topological characteristics of complex networks constructed using the node-wise out-link transition entropy of cross-transition networks as the edge weights have the potential to provide useful indicators for physiological complex networks.

This paper is organized as follows. In Section 2, the implementation process of the proposed method in this study is described in detail. In Section 3, the CPE and the proposed method are used to analyze the unidirectional coupled Honen model with its parameters varied, respectively, and their performance is compared. Next, a realistic EEG dataset recorded during the mental arithmetic task is analyzed by constructing the complex networks using the two methods, respectively, in order to further demonstrate the effectiveness of the proposed method. The discussion and conclusions are given in Sections 4 and 5.

2. Materials and Methods

In this section, the CPE method is briefly introduced, and then the detailed implementation process of the proposed method is described.

2.1. CPE

Based on the permutation entropy and IOTA, Shi et al. proposed the CPE to analyze the information interactions between financial time series [39]. The implementation process is as follows:

1. For two time series with the same length $\{x(t)\}$ and $\{y(t)\}$, $t = [1, 2, \ldots, N]$, their state vectors $X_t = [x_t, x_{t+\tau}, x_{t+2\tau}, \ldots, x_{t+(d-1)\tau}]$ and $Y_t = [y_t, y_{t+\tau}, y_{t+2\tau}, \ldots, y_{t+(d-1)\tau}]$, $t \in [1, 2, \ldots N - (d-1)\tau]$, are obtained through the phase space reconstruction procedure using the delay parameter τ and the embedding dimension d.

2. Performing nondecreasing sort on state vector X_t, and obtaining its position index π_X. Rearranging the state vector Y_t with the position index π_X as the standard, and the result is recorded as $G_t = Y_t(\pi_X)$.

3. Based on the principle of IOTA, the monotonicity is quantified by counting the number of intersection points of the horizontal lines which are drawn from each data point of G_t and G_t itself. The intersections number of the kth state vector is calculated using the following equation:

$$k_t = \sum_{i=1}^{d-2} \sum_{j=i+1}^{d-1} \Theta[(G_t(j+1) - G_t(i))(G_t(i) - G_t(j))] \tag{1}$$

where $\Theta[x]$ is the Heaviside function:

$$\Theta[x] = \begin{cases} 1, & x > 0 \\ 0, & x \leq 0 \end{cases} \tag{2}$$

4. According to this method, all state vectors of the time series are traversed, and the number of the intersections of each state vector can be expressed as a unique integer z, $z \in [0, R]$, $R = (d-1)(d-2)/2$ is the maximum possible number of intersections. For all the $R + 1$ possible values for the integer z_i, $i = 0, 1, \ldots, R$ of intersection points k_t in each state vectors, its probability can be obtained by

$$p(z_i) = \frac{\#\{k_t | k_t = z_i\}}{N - (d-1)\tau} \tag{3}$$

where $1 \leq t \leq N - (d-1)\tau$, $0 \leq i \leq R$, # is the number of elements in the set. Then, after obtaining the probability distribution set $P = \{p(z_i), i = 1, \ldots, R\}$, CPE is defined as:

$$H_{x \to y}(d, \tau) = -\sum_{i=0}^{R} p(z_i) \log_2 p(z_i) \tag{4}$$

According to the above definition, the greater the coupling strength between the two time series, the smaller the CPE. For two random time series, the entropy value reaches the theoretical maximum $\log_2(R + 1)$.

2.2. Cross-Bubble Transition Network (CBTN)

In the original CPE, the process of counting the intersection number of each state vector is essentially a symbolization of it. In the calculation of entropy, only the probability distribution of symbols is considered and the transition behavior between adjacent symbols is ignored. Therefore, the transition network is introduced, in which each symbol is taken as a node and a directional weighted complex network is constructed based on the temporal adjacency of the symbols, with the network weights being the number of transitions between nodes. In addition, to limit the impact of parameter selection on the analysis results, the symbolization process of the bubble entropy was referenced by replacing the intersection number corresponding to each state vector with the number of swaps necessary to sort the state vector in ascending order. The specific implementation process of the cross-bubble transition entropy (Algorithm 1) is as follows:

1. For two equal length time series $\{x(t)\}$ and $\{y(t)\}$, $t = [1, 2, \ldots, N]$, their state vectors $X_t = [x_t, x_{t+\tau}, x_{t+2\tau}, \ldots, x_{t+(d-1)\tau}]$ and $Y_t = [y_t, y_{t+\tau}, y_{t+2\tau}, \ldots, y_{t+(d-1)\tau}]$, $t \in [1, 2, \ldots, N - (d-1)\tau]$, are obtained through the phase space reconstruction procedure using the delay parameter τ and the embedding dimension d. Here, following the parameter choice of bubble entropy, $\tau = 1$;
2. Performing ascending sort on the state vector X_t, and obtaining its position index π_{X_t}. The state vector Y_t was rearranged using the position index π_{X_t} as a criterion and the result was recorded as $G_t = Y_t(\pi_{X_t})$, $t \in [1, 2, \ldots, N - d + 1]$;
3. Sorting the elements in each state vector $G_t = Y_t(\pi_{X_t})$, $t \in [1, 2, \ldots, N - d + 1]$ in ascending order, and calculating the necessary number of swaps S_i, $S_i \in [0, 1, \ldots, d(d-1)/2]$; this is because the number of possible swaps in bubble sort for a d dimensional state vector is from 0 to $d(d-1)/2$;
4. Using S_i, $S_i \in [0, 1, \ldots, d(d-1)/2]$ as network nodes, a directional weighted complex network W was constructed according to the temporal adjacency relationship of S_i and the weight of the network W was the numbers of transition between nodes;

5. In order to reflect the connection relationship between nodes as much as possible, the node-wise out-link transition entropy (NOTE) of the adjacency matrix W was proposed to be used as an indicator parameter. The NOTE was obtained as follows.

The Shannon entropy of each row of the adjacency matrix W was calculated to obtain the local node out-link entropy S_{W_i}, which was used to measure the probability distribution of the output strengths of each node.

$$S_{W_i} = -\sum_{j=0; j \neq i}^{D} w_{ij} \log 2(w_{ij}) \tag{5}$$

where $D = d(d-1)/2$, w_{ij} was the ratio of the output strength from node i to node j to all the output strengths of node i, $\sum_{j=0}^{D} w_{ij} = 1$, and the normalized S_{W_i} was

$$H_{W_i} = S_{W_i}/S_{i,max} \tag{6}$$

where $S_{i,max} = \log 2(D+1)$ was the normalization factor and kept the same for all nodes. The node-wise out-link transition entropy of the adjacency matrix W was

$$H_{NOTE} = \sum_{i=0}^{D} p_i H_{W_i} \tag{7}$$

where p_i was the probability distribution of each node.

The pseudo-code of the proposed algorithm is illustrated as follows.

Algorithm 1. Cross-bubble transition entropy

CBTN $(x(t), y(t), d, \tau)$ // $x(t), y(t)$ are time series. d is embedding dimensions. τ is delay time.
1 performing phase space reconstruction on $x(t), y(t)$ to get
   $X_t = [x_t, x_{t+\tau}, x_{t+2\tau}, \ldots, x_{t+(d-1)\tau}]$ and $Y_t = [y_t, y_{t+\tau}, y_{t+2\tau}, \ldots, y_{t+(d-1)\tau}]$,
   $t \in [1, 2, \ldots, N - (d-1)\tau]$
2 for $t = 1$ to $N - (d-1)\tau$
3   performing ascending sort on X_t to get its position index π_{X_t},
4   Y_t is rearranged according to π_{X_t} to get $G_t = Y_t(\pi_{X_t})$,
5   sorting G_t in ascending order by bubble method and get swaps number S_i,
   $S_i \in [0, 1, \ldots, d(d-1)/2]$.   // $d(d-1)/2$ is the maximum swaps number
6 Using S_i as network nodes, by temporal adjacency relationship of S_i to construct a directed weighted complex network W.
7 for $i = 0$ to D // $D = d(d-1)/2$
8     $S_{W_i} = -\sum_{j=0; j \neq i}^{D} w_{ij} \log 2(w_{ij})$, // $\sum_{j=0}^{D} w_{ij} = 1$
9   normalizing S_{W_i} to get $H_{W_i} = S_{W_i}/S_{i,max}$, // $S_{i,max} = \log 2(D+1)$
10 $H_{NOTE} = 0$.
11 for $i = 0$ to D // $D = d(d-1)/2$
12    $H_{NOTE} = H_{NOTE} + p_i H_{W_i}$.   // p_i is the probability distribution of S_i.
13 return H_{NOTE}.

To demonstrate the performance of the NOTE to track the deterministic dynamical variation in time series, the values with the NOTE and the original CPE were obtained separately for the symbolized sets A = [2 2 4 3 5 1 2] and B = [1 2 2 5 3 2 4]. The probability distributions of the elements in the sets A and B were the same. The probability of individual elements sorted in an ascending order were [0.143, 0.428, 0.143, 0.143, 0.143]. The original CPE method would yield an entropy value of 2.128 for both sets. The directional weighted adjacency matrices constructed for the elements in sets A and B according to their temporal adjacency relationship are shown in Figure 1a,b, respectively. The two adjacency matrices exhibited distinct differences. The NOTE value of these two adjacency matrices was 0.1846 and 0.2925, respectively, which shows the different dynamical variations contained in the sets A and B.

	'1'	'2'	'3'	'4'	'5'
'1'	0	1	0	0	0
'2'	0	1	0	1	0
'3'	0	0	0	0	1
'4'	0	0	1	0	0
'5'	1	0	0	0	0

(a)

	'1'	'2'	'3'	'4'	'5'
'1'	0	1	0	0	0
'2'	0	1	0	1	1
'3'	0	1	0	0	0
'4'	0	0	0	0	0
'5'	0	0	1	0	0

(b)

Figure 1. The directional weighted adjacency matrix constructed from the temporal adjacency relationship for symbol sets A and B. (**a**) The adjacency matrix for the symbolized set A; (**b**) The adjacency matrix for the symbolized set B.

3. Analysis and Results

In this section, in order to verify whether the CBTN can characterize the information interaction between two time series, it was first tested on the unidirectional coupled Honen model and its performance was compared with the original CPE method. Next, the complex network constructed by the CBTN was applied to the realistic EEG recordings during either the resting state or the mental arithmetic task in order to evaluate the performance of the proposed method in detecting the changes in the cognitive load.

3.1. Analysis of Coupled Dynamic Model

The validity of the proposed method was first tested on signals generated using two Henon map unidirectional coupled subsystems with one as the driver subsystem 1 and the other as the responder subsystem 2. The equations of the system are expressed as follows:

$$\begin{cases} x1_{t+1} = 1.4 - x1_t^2 + 0.3 \times y1_t \\ y1_{t+1} = x1_t \\ x2_{t+1} = 1.4 - (C \times x1_t + (1-C) \times x2_t) \times x2_t + 0.3 \times y2_t \\ y2_{t+1} = x2_t \end{cases} \quad (8)$$

The parameter C is a coupling parameter varying from 0 to 1. When C is 0, the two subsystems are entirely independent and there is no definite dynamical behavior between them. When C is 1, the two subsystems are completely synchronized and there is a definite dynamical relationship between them. The values of $x1_1$, $y1_1$, $x2_1$ and $y2_1$ are initialized randomly in the range from 0 to 1. Then, 50,000 points are calculated according to (8) and the first 20,000 are discarded as the transients.

From the definition of CBTN, we can see that the unique parameter relevant to the CBTN is the embedding dimension d. The parameter d defines the embedding spatial dimension of a given time series. Another noteworthy issue is the appropriate signal length in order to obtain reliable results. One fact is that the signal length is limited, and the other is that the calculation process can only be performed in one window. The values of these two parameters determine whether the results of the analysis can be described or not and whether it is possible to extract the deep relationships hidden between the two time series. Here, the determination process of these two parameters is explained by analyzing the unidirectional coupled Honen model with the deterministic coupling relationship. It was selected because the theoretical value of each expected return can be calculated. Based on this theoretical value, it can be evaluated whether the result obtained with a specific parameter setting can converge reliably and stably to the expected value or not. In addition, in order to highlight the impact of the CBTN on the analysis result, the CPE results for the same objects were used as a comparison. When the CPE was used, the embedded dimension was 5 and the delay time was 2.

The appropriate signal length was determined by investigating the effect of the width of the analysis window on the results. For the time series $x1$ and $x2$ obtained using the unidirectional coupled Henon model under a certain coupling strength, the surrogate data $x2^{Surrogate}$ were first calculated by surrogating $x2$ using iAAFT (iterative amplitude-adjusted Fourier transform with five iterations) to mimic the random coupling state. Next, the sliding time window with a fixed moving step of 500 samples was chosen to segment the paired time series $x1$ and $x2$ and $x1$ and $x2^{Surrogate}$. The purpose of using the sliding time window is to enhance the effect of data analysis. The width of the sliding window width was increased from 200 to 5000 samples with a step of 200 samples. With each window width, the coupling strength between $x1$ and $x2^{Surrogate}$ and between $x1$ and $x2$ was calculated using the CBTN for individual sliding windows, obtaining the $H_{NOTE}^{x1-x2^{Surrogate}}$ and H_{NOTE}^{x1-x2}, respectively. The differences $H_{NOTE}^{x1-x2^{Surrogate}} - H_{NOTE}^{x1-x2}$ were first calculated for individual windows and then averaged across windows as the measured difference. The same procedure was repeated for 30 times with each window width and then averaged across repetitions to obtain the average and standard deviation of the measured difference. The coupling strength between $x1$ and $x2$ was set to 0.1, 0.3, and 0.5, respectively, and the results are shown in Figure 2. Figure 2a shows the results of the CBTN method and Figure 2b shows the results of the CPE method. It can be found that the two methods can make a good distinction between different coupling strengths. The CPE method can reach a stable state when the window width is less than 1000 samples, and the CBTN method can reach a stable state when the window width is more than 2000 samples. It was speculated that the CPE was appropriate for the analysis of short time series when it was proposed. Therefore, the appropriate window width for the CBTN is 2000 samples. It should be noted that the differences $H_{CPE}^{x1-x2^{Surrogate}} - H_{CPE}^{x1-x2}$ had a negative value with the CPE method when the coupling strength was 0.1. This is inconsistent with the theory and indicates that the CPE was not capable of differentiating the weak coupling state.

With the determined window width of 2000 samples, the impact of the embedding dimension d on the estimation of the coupling strength was further investigated using the same method as above. The dimension d varied from 3 to 15 with a step of 1. Figure 3 compares the results between the CBTN method and the CPE method. It can be seen that the results of CBTN method tended to be stable with the increase in the embedding dimension d. When the embedding dimension was greater than 10, the measured difference under three coupling strength levels basically reached a stable state. In contrast, the measured difference using the CPE method was greatly affected by the embedded dimension. Within the varying range of the embedded dimension, the measured difference under three coupling strength levels could reach a stable state. When the embedding dimension was greater than 11, the measured difference under the stronger coupling strength (C = 0.3) was even smaller than that under the weaker coupling strength (C = 0.1 and C = 0.2). These results demonstrated that the CBTN method can be less affected by the embedding dimension d compared with the CPE method. More specifically, the embedding dimension would have little influence when it is greater than a certain value. Accordingly, for the unidirectional coupled Henon model, the recommended embedding dimension d for the CBTN method was set to 10.

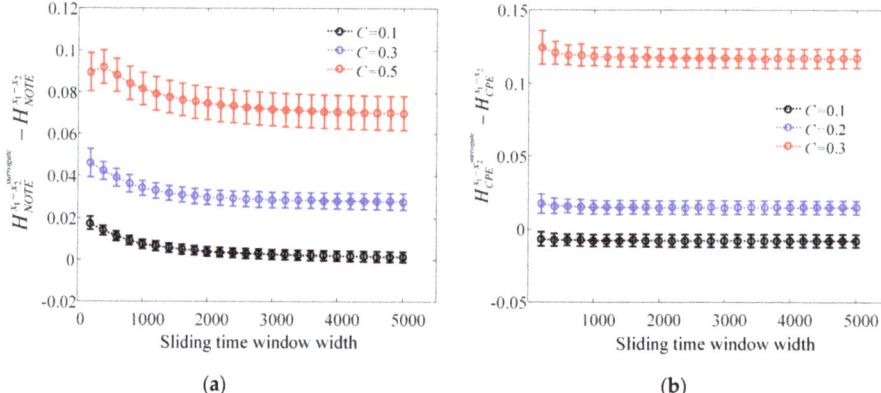

Figure 2. The results of the unidirectional coupled Henon model using CBTN and CPE for coupling analysis, respectively, at different coupling strengths $C = 0.1, 0.3, 0.5$ and when the sliding time window width is varied in steps of 200 samples within $[200, 5000]$. The values of the ordinate are $H_{NOTE}^{x_1-x_2^{surrogate}} - H^{x_1-x_2}$. $x_2^{surrogate}$ can be obtained by surrogating x_2 using the iAAFT method. (**a**) The results of coupling analysis using CBTN (30 repeated calculations); (**b**) The results of coupling analysis using CPE, $d = 5$, $\tau = 2$ (30 repeated calculations).

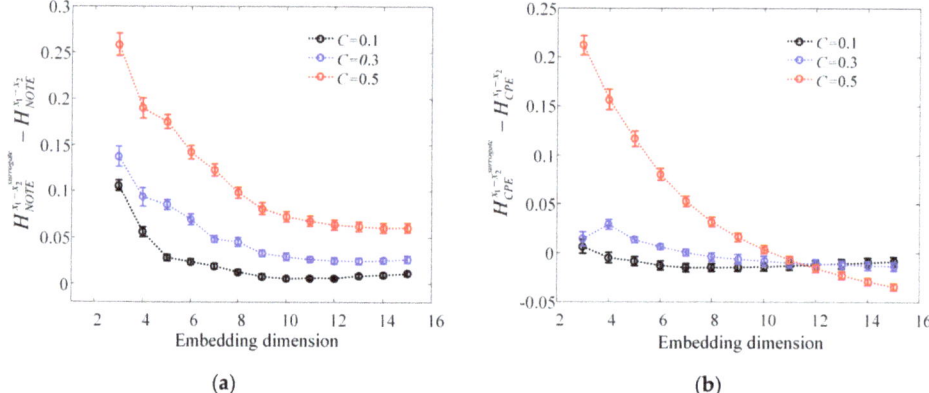

Figure 3. At different coupling strengths $C = 0.1, 0.3, 0.5$, the sliding time window width is fixed at 2000 samples, and the embedding dimension d is taken from 3 to 15; the results of unidirectional coupled Henon model using CBTN and CPE for coupling analysis, respectively. The values of ordinate are $H^{x_1-x_2^{surrogate}} - H^{x_1-x_2}$. $x_2^{surrogate}$ can be obtained by surrogating x_2 using the iAAFT method. (**a**) The results of coupling analysis using CBTN (30 repeated calculations); (**b**) The results of coupling analysis using CPE, $\tau = 2$ (30 repeated calculations).

After the embedding dimension and the sliding window width were determined, the CBTN was used to analyze the unidirectional coupled Honen model under different coupling strengths C. Specifically, the coupling strength C increased from 0 to 0.9 with a step of 0.05. For a given coupling strength, the data of $x1$, $x2$ and $x2^{Surrogate}$ within individual sliding windows were analyzed using the CBTN and CPE methods, respectively, to obtain the H_{NOTE}^{x1-x2}, H_{CPE}^{x1-x2}, $H_{NOTE}^{x1-x2^{Surrogate}}$ and $H_{CPE}^{x1-x2^{Surrogate}}$. The procedure was also repeated 30 times under each coupling strength, and the mean value and the standard deviation across 30 repetitions were calculated. Figure 4 illustrates the average value of H_{NOTE}^{x1-x2}, H_{CPE}^{x1-x2}, $H_{NOTE}^{x1-x2^{Surrogate}}$ and $H_{CPE}^{x1-x2^{Surrogate}}$ across all repetitions under different coupling strength levels. In the figure, the dashed boxes indicated by the arrows are partial zooms of the analysis results. As can be seen from Figure 4, the greatest difference between the analytical results of CBTN and CPE was mainly in the part where the coupling strength was less than 0.2. In this part, the CBTN method gives correct analysis results, while the CPE calculation results are greater than the values under the random coupling state, which is inconsistent with the theory. The possible reason is the CPE method that is based on the probability distribution statistics of symbols cannot distinguish the interactions between time series under weak coupling conditions. In contrast, the proposed CBTN method has good detection capability of interactions between time series with a weak coupling strength.

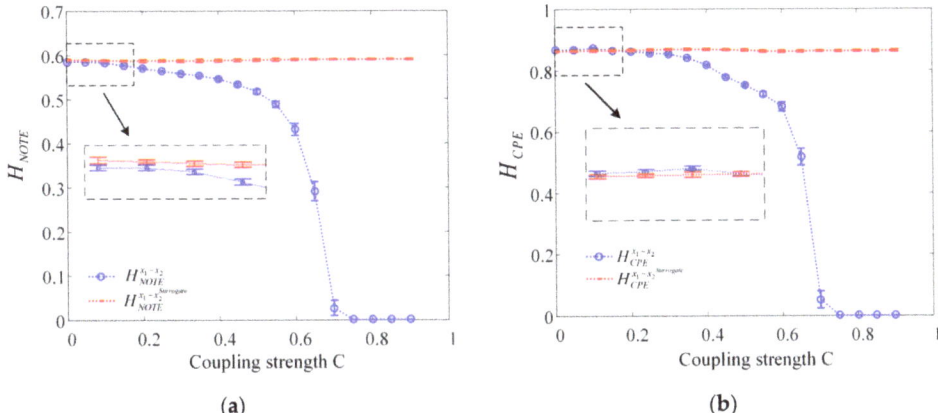

Figure 4. Coupling analysis results for the CBTN-based unidirectional coupled Henon model when the sliding time window is fixed at 2000 samples and the coupling strength C is varied in steps of 0.05 in the range [0, 0.9]. The values of the ordinate are H_{NOTE}. The blue curve is the NOTE between x_1 and x_2 and the red curve is the NOTE between x_1 and $x_2^{surrogate}$. $x_2^{surrogate}$ can be obtained by surrogating x_2 using the iAAFT method (all values in the graph are the result of 30 repeated calculations). (**a**) The results of coupling analysis using CBTN. (**b**) The results of coupling analysis using CPE, $d = 5$, $\tau = 2$.

3.2. Analysis of Realistic EEG in Mental Arithmetic Tasks

In order to demonstrate the performance of the proposed CBTN method on realistic experimental data, the EEG signals of mental arithmetic tasks dataset were used to distinguish the difference between the resting and the arithmetic states of the brain [41]. The dataset can be downloaded freely from the website: https://physionet.org/content/eegmat/1.0.0/, accessed on 23 January 2022. Electrodes were placed according to the international 10/20 scheme and the equipment used was the Neurocom monopolar EEG 23-channel system (Ukraine, XAI-MEDICA). The placement of the silver/silver chloride electrodes on the scalp was prefrontal (Fp1 and Fp2), frontal (F3, F4, Fz, F7, and F8), central (C3, C4, and Cz), parietal (P3, P4, and Pz), occipital (O1 and O2), and temporal (T3, T4, T5, and T6), all referenced to an interconnected ear reference electrode. The impedance between the electrodes and the scalp was less than 5 kΩ, and the sampling rate for each channel was 500 Hz. The acquired EEG signals were filtered using a high-pass filter with a cut-off frequency of 0.5 Hz, a low-pass filter with a cut-off frequency of 45 Hz, and a power line notch filter (50 Hz). The EEG data from 36 subjects (9 males and 27 females, aged 16–26 years) met the requirements for analysis, after a visual inspection of the filtered signals by neuroelectrophysiologists to remove data with poor signal quality. Subject 31 was not included because the length of the recordings was different from that of other subjects. The experiments involved mental arithmetic tasks and each experiment trial was divided into three phases: an adaptation period, a resting state and an arithmetic state. First, the subjects were acclimatized to the experimental conditions for 3 min. Afterwards, the subjects relaxed for 3 min with their eyes closed in the resting state. Finally, the subjects were asked to perform a succession of subtractions in 4 min, each consisting of a four-digit (subtracted number) and two-digit (subtracted number) succession. The two digits were given to the subject verbally and the arithmetic task was not allowed to be performed verbally, but finger movements were allowed. In order to minimize the effect of emotional fluctuations caused by the increased cognitive load of the subjects during intensive cognitive activity on the results of the EEG analysis, the last minute of the resting state and the first minute of the arithmetic state were selected for analysis. Neuroimaging studies showed that the prefrontal and frontal regions were significantly activated during the performance of arithmetic or cognitive tasks [42,43]. Therefore, EEG data collected from seven channels (FP1, FP2, F3, Fz, F4, F7, and F8) in the prefrontal and frontal lobes were used in this study, and the NOTE between any two channels was calculated using the CBTN method to construct an undirected weighted network with the NOTE as the edge weight. Network parameters were extracted from the constructed complex network as a quantitative evaluation indicator of cognitive load. Since the NOTE value is inversely correlated with the coupling strength, in the subsequent analysis, the NOTE values were reversely processed (1 minus the value of NOTE), so that they would adhere to our intuition.

EEG signals were first detrended using the singular value decomposition (SVD) method. Then, the detrended EEG signal was filtered using the harmonic wavelets in the frequency range of 1 to 42 Hz. The obtained resting and arithmetic state EEG signals from seven channels were segmented using a sliding time window with a width of 2000 samples and a moving step of 500 samples. Within each window, the NOTE was estimated between any two of the seven EEG channels using the CBTN method. With the NOTE value as the edge weight, the complex network was constructed and the average clustering coefficient and the global network efficiency of the complex network were calculated. Following the same procedure, all sliding windows were analyzed in turn to obtain the average aggregation coefficient sequence and the global network efficiency sequence of the subject in a state. Since the distribution of the obtained sequences were unknown, the nonparametric permutation test (1000 repeated arrangement sampling) was used to assess the significance between the same sequences in the two states of the subject. The significance level p was set to 0.05. As a comparison, the same operation was performed on this subject using the CPE method with an embedding dimension of 5 and a delay time of 8. The results of the significance analysis between the feature sequences for all the 35 subjects under the two method treatments are shown in Table 1. The values bolded in black in Table 1 indicate statistical insignificance between the two states. As can be seen from the results of the analysis in Table 1, the CBTN method is obviously superior to the CPE method.

In order to confirm whether there were group differences in the EEG signals between the resting and arithmetic states, the mean adjacency matrix of each subject was constructed using the CBTN, and the network parameters of the mean adjacency matrix were extracted for each subject. The same procedure was performed using the CPE as comparison. The EEG data within each sliding window were analyzed using the CBTN to build a complex network, and its adjacency matrix was obtained. All adjacency matrices from the same subject under the same state were averaged. The clustering coefficient and the global network efficiency of the average adjacency matrix were used as a feature for each subject. In this way, the feature in the two states was obtained for individual subjects. The results obtained for all subjects are shown in Figure 5. It can be seen that for most subjects, the mean clustering coefficient of the arithmetic state was smaller than that of the resting state and the global efficiency of the arithmetic state was greater than that of the resting state. This means that the network in the prefrontal area was more efficient and had enhanced information processing capacity during the arithmetic state. It also means an increased cognitive load during the arithmetic state. Figure 6 shows the analysis results of extracting the features of the complex network constructed by the CPE method under two states.

Table 1. Results of nonparametric permutation tests for each subject's feature sequences in the resting and arithmetic states (1000 repeated arrangement sampling, significant level $p = 0.05$).

	CBTN		CPE	
	ACE	GNE	ACE	GNE
Subject 0	0	0	0	0
Subject 1	0	0	0	0
Subject 2	0.004	0.001	0	0
Subject 3	0	0	0	0
Subject 4	**0.973**	**0.540**	**0.665**	0
Subject 5	0	0	**0.518**	**0.342**
Subject 6	0	0	0	0
Subject 7	0	0	0	0
Subject 8	0	0	0	0
Subject 9	0	0	0	0
Subject 10	0	0	0	0
Subject 11	0	0	0	0
Subject 12	0.005	0.018	0	0
Subject 13	0	0	0	0
Subject 14	**0.646**	**0.832**	0	0.018
Subject 15	0	0	0	0
Subject 16	0	0	0	0
Subject 17	**0.125**	**0.158**	**0.398**	0.035
Subject 18	0	0	0	0
Subject 19	0	0	0	0
Subject 20	0	0	0	0
Subject 21	0	0	0	0
Subject 22	0	0	0.001	0.021
Subject 23	0	0	**0.186**	**0.680**
Subject 24	0	0	0.005	0.004
Subject 25	0.011	0.007	**0.483**	**0.603**
Subject 26	0	0	0.049	0.08
Subject 27	0	0	0	0
Subject 28	0	0	0	0
Subject 29	0	0	0	0
Subject 30	0	0	0	0
Subject 32	0	0	**0.591**	**0.895**
Subject 33	0	0	0	0
Subject 34	0	0	0	0
Subject 35	0	0	0	0

ACE (average clustering coefficients), GNE (global network efficiency). Non significant results are shown in bold.

In order to verify whether there was significant difference between the two states at the group level, the results was statistically analyzed using a paired sample t-test. The significance level was set at $p = 0.05$, and statistical analysis was performed on IBM SPSS25.0. The results of the statistical analysis showed that there was a significant difference in the mean clustering coefficients ($p = 0.0013$) and in the global network efficiency ($p = 0.0017$) between the two states using the CBTN method (Figure 5). The results of the statistical analysis also showed that there was a significant difference in the mean clustering coefficients ($p = 0.0056$) and in the global network efficiency ($p = 0.0061$) between the two states using the CPE method. Although both methods can distinguish between the two states, the CBTN analysis was significantly better than the CPE analysis. This suggests that a complex network based on the CBTN using electrodes in the prefrontal and frontal lobe can distinguish well between the two cognitive states, demonstrating the validity of the CBTN method in practical applications.

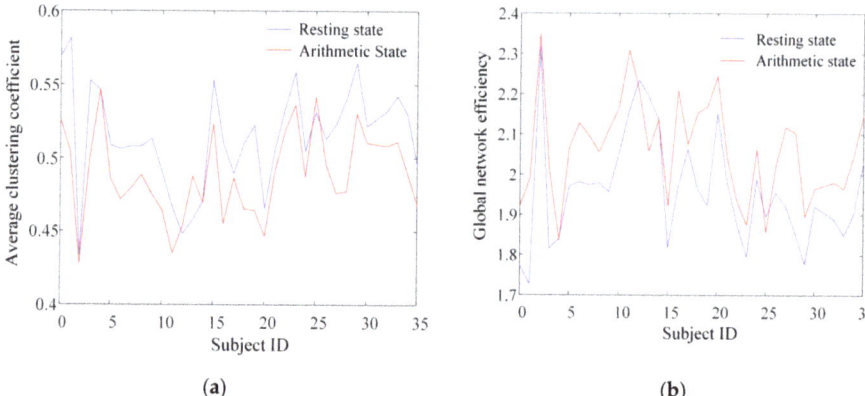

Figure 5. The average clustering coefficient (**a**) and global network efficiency (**b**) of the average adjacency matrix constructed by the CBTN method for each subject.

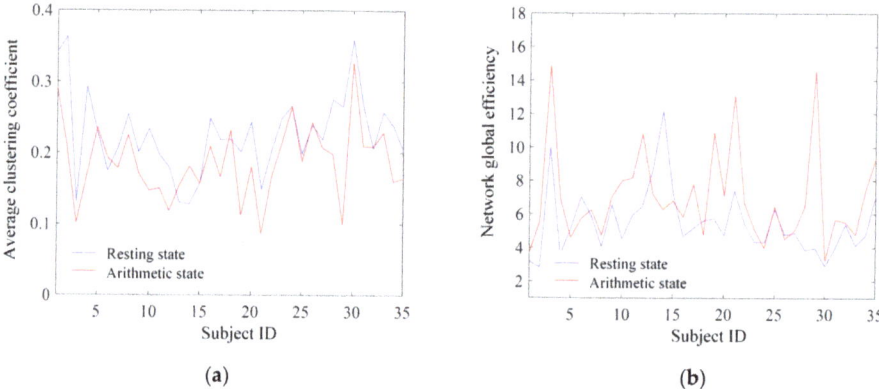

Figure 6. The average clustering coefficient (**a**) and global network efficiency (**b**) of the average adjacency matrix constructed by the CPE method for each subject.

4. Discussion

The aim of this study is to construct a complex network using multichannel EEG signals to enable the assessment of cognitive load. The method of constructing the network directly affects the reliability of the assessment. In the process of using CPE suitable for the analysis of short time series to construct complex networks, it was found that the CPE method suffered from the lack of differentiation ability caused by considering only the probability distribution of symbols and ignoring the transition relationship between symbols in the temporal domain. In addition, as a nonlinear analysis method, the choice of parameters in the CPE had a large impact on the analysis results. In order to alleviate these issues, the CBTN is proposed to measure the coupling relationship between two time series from the perspective of cross-transition networks. The innovation of the CBTN is that it combines the advantages of the transition network and the bubble entropy on the basis of the principle of CPE. The introduction of the transition network solved the problem of ignoring the transition relationship between symbols in the CPE method. The symbolization method with reference to bubble entropy made the analysis result less affected by the embedding dimension. The effectiveness of the method was verified via a comparison with the CPE method on the unidirectional coupled Henon model. Firstly, the results show that the CBTN method could achieve satisfactory results when the signal length reached 2000 samples, although this was slightly larger than that needed in the CPE method. This finding suggests that the CBTN is equally suitable for the analysis of short time series. Secondly, in the experiment study, the result tended to be stable for any coupling strength as long as the embedding dimension of the CBTN was greater than 10. This result indicates that the CBTN method is less affected by the coupling strength. Last, under weak coupling conditions, the CPE method failed to achieve the right results, while the CBTN could still obtain reliable results, indicating that the CBTN was able to uncover the weak coupling relationships between time series. These three properties ensure that the complex network constructed by the CBTN method outperformed the CPE in its ability to analyze cognitive load using EEG datasets. The results of the above analysis clearly demonstrate that the proposed method shows several advantages.

1. This method involves few parameters in use, and the value setting of the parameters has little influence on the analysis results.
2. The cross-transition network allows the method to be more sensitive to weak changes in the information interaction between two time series and is more suitable for analysis in weakly coupled conditions.
3. The normalization measures in the definition of node-wise out-link entropy minimize the impact of intersubject variation on the analysis results.
4. The implementation of the algorithm only involves the ranking of numbers and the probability distribution statistics of symbols, which is easy to be processed and implemented by a computer.

Although the study showed promising results, the limitation of this work should be considered. Firstly, the adjacency matrix of the cross-transition network was a static representation of information interaction between two time series in a period of time. This means that the method was explicitly time-dependent. The analysis of excessively long time series may have caused a reduction in the variation in the adjacency matrix, making identification less effective. This needs further study. Secondly, when using EEG datasets for cognitive load assessment, the electrodes used for analysis were determined subjectively only based on the findings of the neuroimaging, ignoring other aspects of the selection factors. As pointed out in the literature [44], in practical application, the practicality of electrode installation and the comfort of subjects should also be considered. Thirdly, the phase space reconstruction of the time series only considered the influence of the embedded dimension as a variable on the analysis results, and the time delay was set to 1 according to the bubble entropy. In the next research work, the comprehensive impact on the analysis results when these two parameters are variables will be studied in depth.

5. Conclusions

In this study, the advantages of the transition network and the bubble entropy were integrated based on the CPE method, and a new method to measure the coupling strength of two time series was proposed from the perspective of a cross-transition network. It was further used to build complex networks using the multichannel EEG recordings for cognitive load assessment. The results of the unidirectional coupled Honen model showed that this proposed method was not only suitable for the analysis of coupling strength between two short time series, but also had the advantages of being less affected by nonlinear parameters and sensitive to a weak coupling relationship. In addition, the proposed CBTN showed better performance in differentiating cognitive load than the CPE. The new method can be used for state evaluation based on multichannel physiological signals, such as brain state monitoring, quantitative evaluation of various types of mental diseases, and motion decoding based on multichannel electromyography (EMG). It also has an application potential in the financial research field.

Author Contributions: Conceptualization, X.C. and G.X.; methodology, X.C.; software, X.C. and X.Z.; validation, S.Z. and Z.T.; formal analysis, X.C.; investigation, X.C.; resources, S.Z.; data curation, X.C.; writing—original draft preparation, X.C.; writing—review and editing, X.C. and G.X.; visualization, X.Z.; supervision, G.X.; project administration, S.Z.; funding acquisition, S.Z. All authors have read and agreed to the published version of the manuscript.

Funding: This research was funded by the Scientific and Technological Innovation 2030, grant number: 2021ZD0204300 and the Xi'an City Innovation Capability Strengthening Basic Disciplines plan, grant number: 21RGSF0018 and the Key Projects in Shaanxi Province, grant number: 2021GXLH-Z-008).

Institutional Review Board Statement: Not applicable.

Informed Consent Statement: Not applicable.

Data Availability Statement: The raw data supporting the conclusions of this article will be made available by the authors, without undue reservation, to any qualified researcher.

Acknowledgments: We would like to thank all collaborators for their selfless help and guidance in research.

Conflicts of Interest: The authors declare no conflict of interest.

References

1. Sweller, J. Cognitive load during problem solving: Effects on learning. *Cogn. Sci.* **1988**, *12*, 257–285. [CrossRef]
2. Schnotz, W.; Kürschner, C. A reconsideration of cognitive load theory. *Educ. Psychol. Rev.* **2007**, *19*, 469–508. [CrossRef]
3. Paas, F.; Tuovinen, J.E.; Tabbers, H.; Van Gerven, P.W. Cognitive load measurement as a means to advance cognitive load theory. In *Educational Psychologist*; Routledge, Taylor & Francis: London, UK, 2003; pp. 63–71.
4. Useche, S.A.; Cendales, B.; Gómez, V. Measuring fatigue and its associations with job stress, health and traffic accidents in professional drivers: The case of BRT operators. *EC Neurol.* **2017**, *4*, 103–118.
5. Soares, S.M.; Gelmini, S.; Brandao, S.S.; Silva, J. Workplace accidents in Brazil: Analysis of physical and psychosocial stress and health-related factors. *RAM Rev. Adm. Mackenzie* **2018**, *19*. [CrossRef]
6. Burgess, D.J.; Phelan, S.; Workman, M.; Hagel, E.; Nelson, D.B.; Fu, S.S.; Widome, R.; van Ryn, M. The effect of cognitive load and patient race on physicians' decisions to prescribe opioids for chronic low back pain: A randomized trial. *Pain Med.* **2014**, *15*, 965–974. [CrossRef] [PubMed]
7. Hulbert, L.E.; Moisá, S.J. Stress, immunity, and the management of calves. *J. Dairy Sci.* **2016**, *99*, 3199–3216. [CrossRef]
8. Yang, B.; Wang, Y.; Cui, F.; Huang, T.; Sheng, P.; Shi, T.; Huang, C.; Lan, Y.; Huang, Y.-N. Association between insomnia and job stress: A meta-analysis. *Sleep Breath.* **2018**, *22*, 1221–1231. [CrossRef]
9. Heard, J.; Harriott, C.E.; Adams, J.A. A survey of workload assessment algorithms. *IEEE Trans. Hum.-Mach. Syst.* **2018**, *48*, 434–451. [CrossRef]
10. Hart, S.G.; Staveland, L.E. Development of NASA-TLX (Task Load Index): Results of empirical and theoretical research. In *Advances in Psychology*; Elsevier: Amsterdam, The Netherlands, 1988; Volume 52, pp. 139–183.
11. Reid, G.B.; Nygren, T.E. The subjective workload assessment technique: A scaling procedure for measuring mental workload. In *Advances in Psychology*; Elsevier: Amsterdam, The Netherlands, 1988; Volume 52, pp. 185–218.
12. Hart, S.G. NASA-task load index (NASA-TLX); 20 years later. In Proceedings of the Human Factors and Ergonomics Society Annual Meeting, Los Angeles, CA, USA, 16–20 October 2006; pp. 904–908.
13. Arico, P.; Borghini, G.; Di Flumeri, G.; Sciaraffa, N.; Colosimo, A.; Babiloni, F. Passive BCI in operational environments: Insights, recent advances, and future trends. *IEEE Trans. Biomed. Eng.* **2017**, *64*, 1431–1436. [CrossRef]
14. Sibi, S.; Ayaz, H.; Kuhns, D.P.; Sirkin, D.M.; Ju, W. Monitoring driver cognitive load using functional near infrared spectroscopy in partially autonomous cars. In Proceedings of the 2016 IEEE Intelligent Vehicles Symposium (IV), Gothenburg, Sweden, 19–22 June 2016; pp. 419–425.
15. Kadosh, R.C.; Kadosh, K.C.; Linden, D.E.; Gevers, W.; Berger, A.; Henik, A. The brain locus of interaction between number and size: A combined functional magnetic resonance imaging and event-related potential study. *J. Cogn. Neurosci.* **2007**, *19*, 957–970. [CrossRef]
16. Murugesan, S.; Bouchard, K.; Chang, E.; Dougherty, M.; Hamann, B.; Weber, G.H. Hierarchical spatio-temporal visual analysis of cluster evolution in electrocorticography data. In Proceedings of the 7th ACM International Conference on Bioinformatics, Computational Biology, and Health Informatics, Seattle, WA, USA, 2–5 October 2016; pp. 630–639.
17. Antonenko, P.; Paas, F.; Grabner, R.; Van Gog, T. Using electroencephalography to measure cognitive load. *Educ. Psychol. Rev.* **2010**, *22*, 425–438. [CrossRef]
18. Örün, Ö.; Akbulut, Y. Effect of multitasking, physical environment and electroencephalography use on cognitive load and retention. *Comput. Hum. Behav.* **2019**, *92*, 216–229. [CrossRef]

19. Tor, H.T.; Ooi, C.P.; Lim-Ashworth, N.S.; Wei, J.K.E.; Jahmunah, V.; Oh, S.L.; Acharya, U.R.; Fung, D.S.S. Automated detection of conduct disorder and attention deficit hyperactivity disorder using decomposition and nonlinear techniques with EEG signals. *Comput. Methods Programs Biomed.* **2021**, *200*, 105941. [CrossRef] [PubMed]
20. Wiersma, M. Identifying workload levels with a low-cost EEG device using an arithmetic task. In *Faculty of Science and Engineering*; Macquarie University: Sydney, Australia, 2016.
21. Acharya, U.R.; Chua, C.K.; Lim, T.-C.; Dorithy; Suri, J.S. Automatic identification of epileptic EEG signals using nonlinear parameters. *J. Mech. Med. Biol.* **2009**, *9*, 539–553. [CrossRef]
22. Salankar, N.; Qaisar, S.M.; Paweł Pławiak, P.é.; Tadeusiewicz, R.; Hammad, M. EEG based alcoholism detection by oscillatory modes decomposition second order difference plots and machine learning. *Biocybern. Biomed. Eng.* **2022**, *42*, 173–186. [CrossRef]
23. Shahbakhti, M.; Beiramvand, M.; Eigirdas, T.; Solé-Casals, J.; Wierzchon, M.; Broniec-Wójcik, A.; Augustyniak, P.; Marozas, V. Discrimination of Wakefulness from Sleep Stage 1 Using Nonlinear Features of a Single Frontal EEG Channel. *IEEE Sens. J.* **2022**, *22*, 6975–6984. [CrossRef]
24. Paul, J.K.; Iype, T.; Dileep, R.; Hagiwara, Y.; Koh, J.W.; Acharya, U.R. Characterization of fibromyalgia using sleep EEG signals with nonlinear dynamical features. *Comput. Biol. Med.* **2019**, *111*, 103331. [CrossRef]
25. Varela, F.; Lachaux, J.-P.; Rodriguez, E.; Martinerie, J. The brainweb: Phase synchronization and large-scale integration. *Nat. Rev. Neurosci.* **2001**, *2*, 229–239. [CrossRef]
26. Bullmore, E.; Sporns, O. Complex brain networks: Graph theoretical analysis of structural and functional systems. *Nat. Rev. Neurosci.* **2009**, *10*, 186–198. [CrossRef]
27. Shang, J.; Zhang, W.; Xiong, J.; Liu, Q. Cognitive load recognition using multi-channel complex network method. In Proceedings of the International Symposium on Neural Networks, Sapporo, Hakodate, Muroran, Japan, 21–26 June 2017; pp. 466–474.
28. Kakkos, I.; Dimitrakopoulos, G.N.; Gao, L.; Zhang, Y.; Qi, P.; Matsopoulos, G.K.; Thakor, N.; Bezerianos, A.; Sun, Y. Mental workload drives different reorganizations of functional cortical connectivity between 2D and 3D simulated flight experiments. *IEEE Trans. Neural Syst. Rehabil. Eng.* **2019**, *27*, 1704–1713. [CrossRef]
29. Shovon, M.; Islam, H.; Nandagopal, N.; Vijayalakshmi, R.; Du, J.T.; Cocks, B. Directed connectivity analysis of functional brain networks during cognitive activity using transfer entropy. *Neural Process. Lett.* **2017**, *45*, 807–824. [CrossRef]
30. Suresh, K.; Ramasamy, V.; Daniel, R.; Chandra, S. Characterizing EEG Electrodes in Directed Functional Brain Networks Using Normalized Transfer Entropy and PageRank. In *Handbook of Artificial Intelligence in Healthcare*; Springer: Berlin/Heidelberg, Germany, 2022; pp. 27–49.
31. Wang, J.; Wang, L.; Zang, Y.; Yang, H.; Tang, H.; Gong, Q.; Chen, Z.; Zhu, C.; He, Y. Parcellation-dependent small-world brain functional networks: A resting-state fMRI study. *Hum. Brain Mapp.* **2009**, *30*, 1511–1523. [CrossRef] [PubMed]
32. Steuer, R.; Kurths, K., Jr.; Daub, C.O.; Weise, J.; Selbig, J. The mutual information: Detecting and evaluating dependencies between variables. *Bioinformatics* **2002**, *18*, S231–S240. [CrossRef] [PubMed]
33. Vejmelka, M.; Paluš, M. Inferring the directionality of coupling with conditional mutual information. *Phys. Rev. E* **2008**, *77*, 026214. [CrossRef]
34. Schreiber, T. Measuring information transfer. *Phys. Rev. Lett.* **2000**, *85*, 461. [CrossRef]
35. Hempel, S.; Koseska, A.; Kurths, K., Jr.; Nikoloski, Z. Inner composition alignment for inferring directed networks from short time series. *Phys. Rev. Lett.* **2011**, *107*, 054101. [CrossRef]
36. Liu, L.-Z.; Qian, X.-Y.; Lu, H.-Y. Cross-sample entropy of foreign exchange time series. *Phys. A Stat. Mech. Its Appl.* **2010**, *389*, 4785–4792. [CrossRef]
37. Shu, Y.; Zhao, J. Data-driven causal inference based on a modified transfer entropy. *Comput. Chem. Eng.* **2013**, *57*, 173–180. [CrossRef]
38. Kiwata, H. Relationship between Schreiber's transfer entropy and Liang-Kleeman information flow from the perspective of stochastic thermodynamics. *Phys. Rev. E* **2022**, *105*, 044130. [CrossRef]
39. Shi, W.; Shang, P.; Lin, A. The coupling analysis of stock market indices based on cross-permutation entropy. *Nonlinear Dyn.* **2015**, *79*, 2439–2447. [CrossRef]
40. Manis, G.; Aktaruzzaman, M.D.; Sassi, R. Bubble Entropy: An Entropy almost Free of Parameters. *IEEE Trans. Bio-Med. Eng.* **2017**, *64*, 2711–2718.
41. Zyma, I.; Tukaev, S.; Seleznov, I.; Kiyono, K.; Popov, A.; Chernykh, M.; Shpenkov, O. Electroencephalograms during mental arithmetic task performance. *Data* **2019**, *4*, 14. [CrossRef]
42. Yu, J.; Pan, Y.; Ang, K.K.; Guan, C.; Leamy, D.J. Prefrontal cortical activation during arithmetic processing differentiated by cultures: A preliminary fNIRS study. In Proceedings of the 2012 Annual International Conference of the IEEE Engineering in Medicine and Biology Society, San Diego, CA, USA, 28 August–1 September 2012; pp. 4716–4719.
43. Menon, V.; Mackenzie, K.; Rivera, S.M.; Reiss, A.L. Prefrontal cortex involvement in processing incorrect arithmetic equations: Evidence from event-related fMRI. *Hum. Brain Mapp.* **2002**, *16*, 119–130. [CrossRef] [PubMed]
44. Shahbakhti, M.; Beiramvand, M.; Rejer, I.; Augustyniak, P.; Broniec-Wojcik, A.; Wierzchon, M.; Marozas, V. Simultaneous eye blink characterization and elimination from low-channel prefrontal EEG signals enhances driver drowsiness detection. *IEEE J. Biomed. Health Inform.* **2021**, *26*, 1001–1012. [CrossRef]

Article

Effects of Sleep Deprivation on the Brain Electrical Activity in Mice

Alexey N. Pavlov [1,*], Alexander I. Dubrovskii [2], Olga N. Pavlova [2] and Oxana V. Semyachkina-Glushkovskaya [3]

1. Faculty of Nonlinear Processes, Saratov State University, Astrakhanskaya Str. 83, 410012 Saratov, Russia
2. Physics Department, Saratov State University, Astrakhanskaya Str. 83, 410012 Saratov, Russia; paskalkamal@mail.ru (A.I.D.); pavlov_lesha@yahoo.com (O.N.P.)
3. Biology Department, Saratov State University, Astrakhanskaya Str. 83, 410012 Saratov, Russia; glushkovskaya@mail.ru
* Correspondence: pavlov.alexeyn@gmail.com

Abstract: Sleep plays a crucial role in maintaining brain health. Insufficient sleep leads to an enhanced permeability of the blood–brain barrier and the development of diseases of small cerebral vessels. In this study, we discuss the possibility of detecting changes in the electrical activity of the brain associated with sleep deficit, using an extended detrended fluctuation analysis (EDFA). We apply this approach to electroencephalograms (EEG) in mice to identify signs of changes that can be caused by short-term sleep deprivation (SD). Although the SD effect is usually subject-dependent, analysis of a group of animals shows the appearance of a pronounced decrease in EDFA scaling exponents, describing power-law correlations and the impact of nonstationarity as a fairly typical response. Using EDFA, we revealed an SD effect in 9 out of 10 mice (Mann–Whitney test, $p < 0.05$) that outperforms the DFA results (7 out of 10 mice). This tool may be a promising method for quantifying SD-induced pathological changes in the brain.

Keywords: detrended fluctuation analysis; long-range correlations; electroencephalogram; sleep deprivation; nonstationarity

1. Introduction

Sleep plays a critical role in maintaining the health of the central nervous system and resisting small vessel disease in the brain. Over the past decades, there has been a better understanding of the effects of sleep on the body [1–4]. Sleep is important for attention and learning and affects long-term memory, decision-making, etc. [5,6]. It is vital to maintain good overall brain health, and prolonged periods of the absence of sleep can have serious consequences. Good sleep reduces the risk of neurodegenerative disorders, and insufficient sleep leads to sterile inflammation in the absence of infection [7–9] and an enhanced permeability of the blood–brain barrier (BBB) [8,10]. Total sleep deprivation (SD) of rats resulted in their death [11]. In humans, the longest wakefulness time (11 days) is accompanied by hallucinations and various cognitive impairments [12]. Thus, it seems clear that sleep plays an important role in restoring brain function. Sleep is a biomarker of BBB permeability, and electroencephalography (EEG) is an important informative platform for analyzing BBB leakage, especially in amyloid lesions of small vessels of the brain [13]. It is interesting to note that the opening of the BBB and deep sleep are accompanied by similar activation of toxins clearance from the brain [13]. Thus, nighttime EEG patterns also hide information about lymphatic drainage and cleansing functions of the brain. Detecting such information requires techniques that deal with nonstationary signal processing, and one such tool is the detrended fluctuation analysis (DFA).

Since its appearance [14,15], DFA has attracted considerable attention in many areas of research, where correlation features of experimental datasets are used to characterize the complex dynamics of natural systems [16–22]. The traditional correlation function $C(\tau)$

has two main restrictions, the first of which is the decay of $C(\tau)$ with increasing time lag τ, which is fast for broadband random processes. The latter limits the ability to compute the scaling exponent describing long-range power-law correlations, because $C(\tau)$ approaches zero and becomes comparable to computational errors for noisy datasets. The second restriction arises for time-varying dynamics, when the value of C is determined by two time moments t_1, t_2. Only for stationary processes there is a dependence on their difference $\tau = t_2 - t_1$, and the correlation function is described by one variable. Many natural processes do not satisfy this requirement, and the traditional approach is used under the assumption of quasistationarity for short segments of the dataset or after excluding the trend due to data filtering. The origin of nonstationarity differs. It can be caused by recording equipment failures or by transients between various system states. Otherwise, it appears due to internal slow dynamics with time scales comparable to the duration of the available datasets. In the latter case, we interpret part of the internal dynamics of a system with time-varying components as a trend. The advantage of DFA is the inclusion of data filtering (detrending) in the signal processing algorithm [14]. Moreover, this detrending is carried out for each time scale separately, which is important for inhomogeneous datasets. Another advantage is the transformation of the decreasing correlation function into a growing dependence of the root mean square (RMS) fluctuations of the signal profile around the local trend on the time scale, and the scaling exponent describing its power-law features is easier to estimate, especially in the region of long-range correlations [15]. The DFA has some limitations that were discussed in earlier studies [23–26]. Despite the detrending procedure, nonstationarity influences the results, and data preprocessing is still important for analyzing complex systems using experimentally recorded signals [27].

In its original version [14,15], the DFA considers one basic type of nonstationarity, namely, slow variations in the local mean value (trend). However, natural processes can include other types of time-varying behavior, e.g., repeated regular or random switching between system states, variations in energy, etc. The application of DFA can lead to misinterpretation of scaling exponents for inhomogeneous datasets, where segments with small and large RMS fluctuations can coexist, and their number affects the results. Several attempts have been made to modify the conventional method, such as multifractal DFA, which introduces a number of quantities instead of a single scaling exponent [28,29]. Recently, we proposed another modification that takes into account local RMS fluctuations and estimates two scaling exponents describing the features of power-law correlations and the impact of nonstationarity [30]. This approach, extended DFA (EDFA) [31,32], has been applied to various types of physiological processes to improve the diagnostic capabilities of the conventional method. The main idea of EDFA is to take into account the difference between the maximum and minimum local RMS fluctuations of the signal profile (random walk) around the trend depending on the time scale. Here, we perform some modification of the EDFA to provide a more stable computation algorithm, and consider the standard deviation of the local RMS fluctuations. Such improvement allows us to avoid or at least reduce the effect of artifacts in experimental measurements, when localized artifacts or short-term instabilities strongly influence the RMS fluctuations within the conventional DFA and alter the quantitative measures of long-range correlations.

To illustrate the EDFA's ability to characterize effects of SD on the brain electrical activity, here we analyze EEGs acquired in awake mice in two different states—background electrical brain activity and activity after SD [33–37], when the animals did not sleep for a day. Unlike prolonged SD, the effects of short-term SD are less obvious. Here, we study how one-day SD alters long-range power-law correlations in electrical activity in the brain. The manuscript is organized as follows. In Section 2, we describe the subjects, experimental procedures, and data measurements used in this work. We also provide a brief description of DFA and its modified version, EDFA. The results of EEG studies in mice during background activity and after sleep deprivation are presented in Section 3. Section 4 summarizes the main findings of the study.

2. Methods and Experiments

2.1. DFA

DFA is a variant of the correlation analysis of experimental datasets originally proposed by Peng et al. [14,15]. It is based on RMS analysis of signal profile and includes the following steps:

(1) Transition from signal $x(i), i = 1, \ldots, N$ to its profile $y(k), k = 1, \ldots, N$

$$y(k) = \sum_{i=1}^{k}[x(i) - \langle x \rangle], \quad \langle x \rangle = \sum_{i=1}^{N} x(i). \tag{1}$$

(2) Segmentation of the profile $y(k)$ into parts of length n ($n<N$).

(3) Computation of the local trend $y_n(k)$ for each segment using a least-squares straight-line fit.

(4) Estimation of the standard deviation,

$$F(n) = \sqrt{\frac{1}{N}\sum_{k=1}^{N}[y(k) - y_n(k)]^2}. \tag{2}$$

(5) Implementation steps 2–4 over a wide range of n.

(6) Computation of the scaling exponent α,

$$F(n) \sim n^{\alpha}. \tag{3}$$

Power-law dependence (3) is observed for various random processes, but many real-world datasets with an inhomogeneous structure often exhibit different local slopes of $\lg F$ vs. $\lg n$, and α may differ for short-range and long-range correlations. DFA is usually applied to reveal the features of complex dynamics related to the region of long-range correlations. Specific values of α, associated with $\alpha < 0.5$, $\alpha = 0.5$, and $0.5 < \alpha < 1$, describe, respectively, anti-correlated behavior (alternation of large and small values of $x(i)$), lack of correlations (e.g., white noise), and positive power-law correlations (large values of $x(i)$ tend to follow large values, and vice versa) [15]. Positive correlations, which may differ from power-law behavior, are associated with $\alpha > 1$.

2.2. EDFA

Signal properties can vary strongly between different parts of a dataset. This is observed, e.g., for transients from one state to another, when well-pronounced variations in the local mean value affect the scaling exponent. Considering datasets with and without such transients can lead to distinct results of DFA. In a recent study [27], we illustrated the effects of nonstationarity for several cases: low-frequency trend, intermittent dynamics, and nonstationarity in energy. Besides the case when time-varying dynamics occurs throughout the signal, the α exponent is also influenced by short-term failures of the recording equipment or artifacts. Such data segments provide distinct local standard deviations (2) compared to the averaged quantities. In order to characterize the differences in nonstationarity across the entire signal, we have proposed the following modification of the method, called EDFA [31,32]. Within this approach, a new measure

$$dF(n) = \max[F_{loc}(n)] - \min[F_{loc}(n)], \quad F_{loc}(n) = \sqrt{\frac{1}{n}\sum_{k=1}^{n}[y(k) - y_n(k)]^2} \tag{4}$$

is introduced, where $F_{loc}(n)$ are the local standard deviations of the profile from the trend, which are estimated for each segment. The difference $dF(n)$ contains information about the impact of signal inhomogeneity. If the properties of the signal vary insignificantly, $dF(n)$

takes values close to zero. Otherwise, a wide distribution of $F_{loc}(n)$ appears, and $dF(n)$ varies with n, exhibiting power-law behavior characterized by the scaling exponent β

$$dF(n) \sim n^\beta. \tag{5}$$

In this definition, β becomes highly sensitive to artifacts in experimental recordings. The existence of a single artifact can lead to a large $F_{loc}(n)$ associated with $\max[F_{loc}(n)]$, and the latter reduces the stability of the EDFA method. In particular, the $dF(n)$ dependence can show strong fluctuations with n. A more stable algorithm is based on the statistical analysis of $F_{loc}(n)$, and the use of the standard deviation $\sigma(F_{loc}(n))$ as a measure of the signal inhomogeneity. Thus, here we propose to consider the dependence

$$\sigma(F_{loc}(n)) \sim n^\beta. \tag{6}$$

Figure 1 shows both dependences (5) and (6) in a lg–lg plot for the case of $1/f$-noise used as a simple example of a homogeneous process with power-law correlations. This figure confirms that the latter definition provides reduced variability in the estimated values. Thus, standard error of the β estimates decreases from 0.0038 for the definiton (5) to 0.0023 for the definiton (6).

For physiological datasets, differences are usually larger. Although the exponents β in Equations (5) and (6) may differ, we use the same designation (β) to quantify the impact of nonstationarity and a more stable algorithm based on $\sigma(F_{loc}(n))$ for its evaluation. The β exponent can take as positive, as negative values [30]. Both α and β exponents describe different signal properties and are independent quantities.

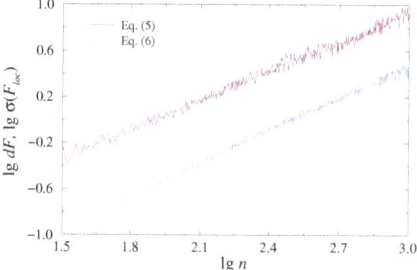

Figure 1. Dependences described by Equations (5) and (6) in the lg–lg plot for $1/f$-noise. The β-exponent is estimated with the standard errors 0.0038 and 0.0023, respectively.

2.3. Subjects and Experiments

Experiments were carried out on ten C57BL/6 male mice (20–25 g) in accordance with the Guide for the Care and Use of Laboratory Animals (8th ed., The National Academies Press, Washington, 2011). The protocols were approved by the Local Bioethics Commission of the Saratov State University. The mice were kept in a light/dark environment with the lights on from 8:00 to 20:00 and fed ad libitum with standard rodent food and water. The ambient temperature and humidity were maintained at 24.5 ± 0.5 °C and 40–60%, respectively.

A two-channel cortical EEG (Pinnacle Technology, Taiwan) was recorded (Figure 2) using two silver electrodes (tip diameter 2–3 μm) located at a depth of 150 μm in coordinates (L: 2.5 mm and D: 2 mm) from Bregma on either sides of the midline under inhalation anesthesia with 2% isoflurane at 1 L/min N_2O/O_2—70:30. The head plate was mounted and small burr holes were drilled. Thereafter, wire EEG leads were inserted into burr holes on one side of the midline between the skull and the underlying dura mater. EEG leads were fixed with dental acrylic. Ibuprofen (15 mg/kg) for relief of postoperative pain was provided in water supply for two to three days before surgery and for three or more days after surgery. Before starting the experiment, the mice were given 10 days to recover from surgery.

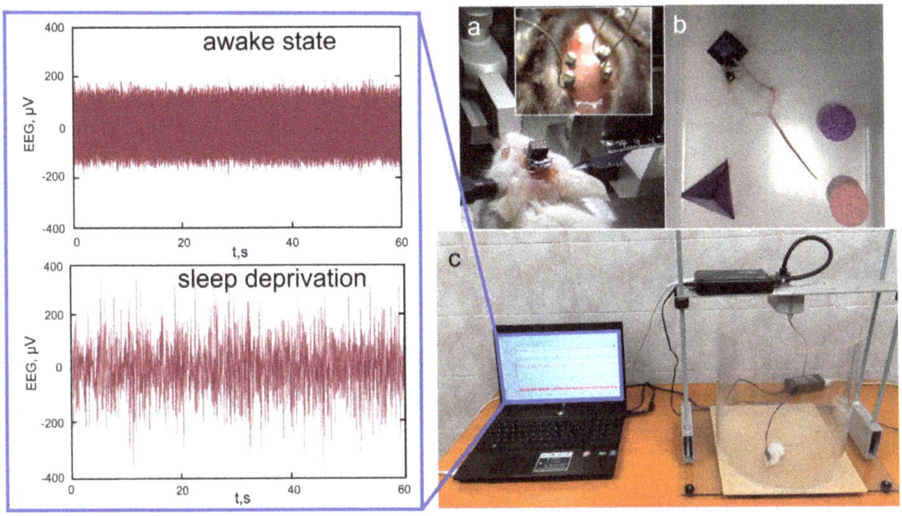

Figure 2. Experiment design: (**a**) implantation of a two–channel cortical EEG, (**b**) SD by presenting new objects to the mouse, and (**c**) EEG recording in an awake mouse. Insert shows experimental EEG signals—voltages (µV) vs. time (seconds).

As standard sleep staging rules for mice are not currently available, we referred to the visual assessment criteria from the studies [1,38]. Sleep deprivation was carried out according to the method described in [39], with adaptation to the vivarium regime. The mice were deprived of sleep from 8:00 pm to 8:00 am and were immediately used for the experiment. Sleep deprivation was maintained by bringing new objects and sounds into the experiment room [40]. The mice were constantly monitored to make sure they were actually studying objects.

Signals were measured in awake and sleeping mice (day 1, 10-h recording), and after SD (day 2, 4-h recording). This study compares two states: (1) awake mice, background EEG activity, and (2) awake mice after SD. All recordings were done with a sampling rate of 2 kHz. At the stage of preprocessing, twelve 5-min segments with a quite homogeneous structure and less distorted by artifacts were selected for each EEG channel. Every 5-min segment was analyzed with EDFA to evaluate $\sigma(F_{loc}(n))$ and the β exponent. The results for each mouse state were averaged over all selected segments and both channels.

3. Results and Discussion

According to earlier studies [31,32], the slopes of the dependences (3) and (5) on the lg–lg plot vary with the segment length, which characterizes the range of power-law correlations. Often, significant differences between the properties of long-range and short-range correlations are observed in many types of physiological signals, which was demonstrated in pioneering works [14,15]. Knowledge of the features of power-law dependences (3) and (5) is important for establishing informative markers that quantify changes caused by transitions between different physiological states. In the case of the rat EEG, noticeable changes in correlation properties during sleep were found in the region of slow-wave dynamics associated with the $\lg n > 3.3$ range, which refers to frequencies below 1 Hz. To specify this range for the current study, we performed a preliminary visual analysis of the EDFA results. By analogy with the works in [31,32], transitions between distinct physiological states (dynamics before or after SD) can be observed in the area of long-range correlations, but changes between slopes, quantitatively determined by the exponents α

and β, are usually subject-dependent and may also vary throughout a single recording for the same animal. In some cases, the range $\lg n > 3.3$ is appropriate to illustrate the effect of sleep deprivation. In other cases, lower frequencies associated with larger values of $\lg n$ seem preferable. Thus, Figure 3 shows two examples of the dependence $\lg F$ vs. $\lg n$ for individual 5-min EEG segments measured in awake mice before and after SD. They illustrate the most pronounced differences between the states for $\lg n > 3.9$. As $\lg n$ decreases, the distinctions in slopes are still observed, but they become weaker. Insert show the results of statistical analysis performed to select an appropriate range of scales (the values of $\lg n$ related to changes in the slopes of $\lg F$ vs. $\lg n$).

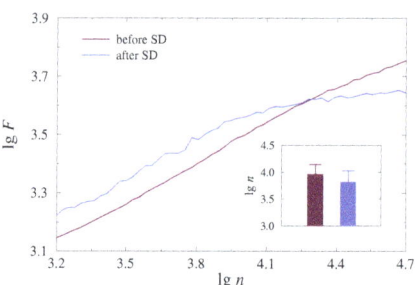

Figure 3. Examples of the dependence (3) in the lg–lg plot for 5-min EEG segments acquired in an awake mice before and after SD. Insert shows the results of statistical analysis over different EEG segments.

For the β-exponent, the effect can be more pronounced, as this exponent can change its sign upon transition to another physiological state. This is illustrated in Figure 4 for the same measurements as in Figure 3. Again, the range $\lg n > 3.9$ is better suited for quantifying the distinctions caused by SD. Analogous visual analysis performed on other animals or data segments allowed us to capture this range of scales for further statistical analysis of the groups.

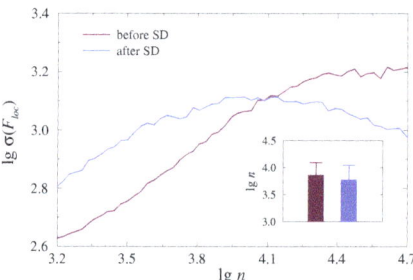

Figure 4. Examples of the dependence (6) in the lg–lg plot for 5-min EEG segments acquired in an awake mice before and after SD. Insert shows the results of statistical analysis over different EEG segments.

When conducting statistical analysis, several important points should be mentioned. First, there is a significant variability in the measures α, β within each state, due to the individual peculiarities of animals. The latter complicates comparison of the states based on absolute values of the scaling exponents, and accounting for differences between exponents α_1, β_1 related to state 1 (awake mice before SD) and α_2, β_2 associated with state 2 (awake mice after SD) seems to be a more promising approach. Thus, we introduce two measures

$$\Delta \alpha = \alpha_1 - \alpha_2, \quad \Delta \beta = \beta_1 - \beta_2 \tag{7}$$

for a quantitative description of SD effects.

Another circumstance is the significant variability in scaling exponents between different parts of each recording. On the one hand, we can take longer datasets (e.g., several hours), estimate the corresponding values of α and β, and then compare these quantities for the two states under consideration. However, this way is accompanied by time-varying dynamics and several types of nonstationary behavior that can change the expected values of the scaling exponents. On the other hand, we can select fairly homogeneous (more stationary) segments, analyze them with EDFA, and then average the results for each animal and each condition. Our preliminary analysis of the simulated datasets [27] showed that this way gives more stable and reliable estimations, and we use it here for EEG processing.

The established distinctions caused by SD for the entire group of animals are given in Figure 5, where different symbols indicate distinct responses, and in Table 1. For six out of 10 mice, a pronounced effect of SD is observed, characterized by a decrease in the α and β exponents, i.e., positive values of $\Delta\alpha$ and $\Delta\beta$ (Figure 5, circles). Several animals (three out of 10 mice) demonstrated relatively subtle signs of SD-induced changes (Figure 5, triangles), although these changes are significant according to the Mann–Whitney test ($p < 0.05$). In this study, only one day of sleep deprivation was used. Longer SD periods are expected to elicit stronger responses; however, our goal was to examine the effects of short-term SD when sleep deprivation is not associated with neurodegenerative disorders. According to Figure 5, one mouse showed a different reaction (square), but this behavior can be treated atypical compared to other animals. Our results indicate that short-term effects of SD can be detected in EEG recordings, although the strength of the response is subject-dependent. Moreover, accounting for the β exponent of the proposed EDFA can surpass the α exponent of the standard DFA—the $\Delta\beta$ range is about twice as large as the $\Delta\alpha$ range (0.12 ± 0.04 versus 0.05 ± 0.02). Consequently, the changes in the features of nonstationarity caused by SD are more pronounced than the changes in the properties of long-range correlations associated with them.

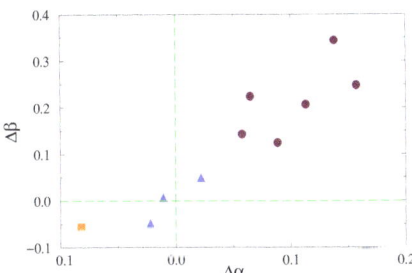

Figure 5. Individual responses of mice from the entire group to one-day SD quantified with the differences between the scaling exponents (7) before and after SD ($\Delta\alpha$ and $\Delta\beta$). Maroon circles show pronounced effects of SD, blue triangles indicate relatively subtle signs of SD-induced changes, and an orange square marks atypical response.

Thus, in this study we show that EDFA sensitively reflects the changes induced by SD. The sleep is a natural factor of activation of the lymphatic clearing and drainage functions of the brain [1,13]. The SD causes significant suppression of the clearance of toxins from the brain [1]. There are animal data suggesting that sleep deficit leads to sterile inflammation [7–9], an increase in the BBB permeability [8,10], and long SD is accompanied by hallucination and various cognitive deficiencies [12]. We hypothesized that sleep is a biomarker of the BBB permeability and EEG is an important informative platform for the analysis of BBB leakage and the cerebral lymphatic functions [20]. We show that EDFA may be applied to study changes in the electrical brain activity after SD.

Table 1. Characterization of SD effects with measures (7). The results are given as mean values ± SE. Asterisks indicate statistically significant changes according to the Mann–Whitney test ($p < 0.05$). EDFA shows significant changes for 9 out of 10 animals ($\Delta\beta$), while DFA provides significant distinctions for 7 out of 10 mice ($\Delta\alpha$). The last column indicates that changes in $\Delta\beta$ are stronger in 8 mice ($|\Delta\beta|>|\Delta\alpha|$).

Experiment	DFA ($\Delta\alpha$)	EDFA ($\Delta\beta$)	$|\Delta\beta|/|\Delta\alpha|$
1	0.14 ± 0.02 *	0.34 ± 0.03 *	2.4
2	0.06 ± 0.02 *	0.22 ± 0.02 *	3.7
3	−0.01 ± 0.01	0.01 ± 0.01	1.0
4	0.06 ± 0.02 *	0.14 ± 0.02 *	2.3
5	0.11 ± 0.02 *	0.21 ± 0.02 *	1.9
6	0.02 ± 0.01	0.05 ± 0.01 *	2.5
7	−0.03 ± 0.01	−0.04 ± 0.01 *	1.3
8	0.09 ± 0.02 *	0.13 ± 0.02 *	1.4
9	−0.08 ± 0.02*	−0.06 ± 0.02 *	0.8
10	0.16 ± 0.03 *	0.25 ± 0.04 *	1.6

4. Conclusions

The scaling features of long-range power-law correlations are important markers of the complex dynamics of many real-world systems that can be used to diagnose the state of a system based on experimentally measured datasets. We have discussed a modified approach to the fluctuation analysis of inhomogeneous processes, in which the nonstationarity varies over the entire signal. This approach, EDFA, computes two scaling exponents that quantify power-law correlations and nonstationarity features. To provide a more stable computational procedure and reduce the impact of localized artifacts, we consider the standard deviation of the local RMS fluctuations of the signal profile around the trend, rather than the difference of extreme values. The benefits of the latter procedure are illustrated using simulated datasets.

We then applied this approach to EEG signals in mice to reveal signs of changes in electrical activity of the brain that could be caused by a day's sleep deprivation. These signs can be fairly subtle, in contrast to the effects of prolonged SD, taking into account the significant variability of characteristics during long-term EEG recordings. Using a group of 10 mice, we found quite strong reductions in α and β scaling exponents in six animals, with only one mouse showing a pronounced opposite effect. In these animals, the β exponent provided stronger responses than the α exponent of the conventional DFA. Thus, the proposed modified version of the method can be a useful prognostic tool for the evaluation of SD-mediated suppression of the clearance of toxins from the brain that is in accordance with the work [1]. Important open questions that could be further analyzed are the role of SD duration and the factor of age.

Author Contributions: Conceptualization, A.N.P. and O.V.S.-G.; methodology, A.N.P.; data curation, A.I.D.; software, O.N.P.; formal analysis, A.I.D., O.N.P.; investigation, A.N.P., O.N.P.; writing—original draft preparation, A.N.P. and O.V.S.-G.; writing—review and editing, A.N.P. and O.V.S.-G.; visualization, A.N.P.; supervision, A.N.P. and O.V.S.-G.; project administration, O.V.S.-G.; funding acquisition, O.V.S-G. All authors have read and agreed to the published version of the manuscript.

Funding: This work was supported by the grant of the Government of the Russian Federation No. 075-15-2019-1885 in the part of experimental studies and numerical analysis. A.P. acknowledges the support by the grant of the Russian Science Foundation (Agreement 19-12-00037) in the part of theoretical studies (the development of EDFA method).

Institutional Review Board Statement: The study was carried out in accordance with the Guide for the Care and Use of Laboratory Animals (8th ed., The National Academies Press, Washington, 2011), and approved by the Local Bioethics Commission of the Saratov State University (protocol No. 12, 17.02.2020).

Informed Consent Statement: Not applicable.

Data Availability Statement: The data that support the findings of this study are available from the corresponding author upon reasonable request.

Conflicts of Interest: The authors declare no conflict of interest.

Abbreviations

The following abbreviations are used in this manuscript:

DFA	Detrended fluctuation analysis
EDFA	Extended detrended fluctuation analysis
EEG	Electroencephalogram
SD	Sleep deprivation
RMS	Root mean square

References

1. Xie, L.; Kang, H.; Xu, Q.; Chen, M.J.; Liao, Y.; Thiyagarajan, M.; O'Donnell, J.; Christensen, D.J.; Nicholson, C.; Iliff, J.J.; et al. Sleep drives metabolite clearance from the adult brain. *Science* **2013**, *342*, 373–377. [CrossRef]
2. Depner, C.M.; Stothard, E.R.; Wright, K.P., Jr. Metabolic consequences of sleep and circadian disorders. *Curr. Diabetes Rep.* **2014**, *14*, 507. [CrossRef]
3. Fultz, N.E.; Bonmassar, G.; Setsompop, K.; Stickgold, R.A.; Rosen, B.R.; Polimeni, J.R.; Lewis, L.D. Coupled electrophysiological, hemodynamic, and cerebrospinal fluid oscillations in human sleep. *Science* **2019**, *366*, 628–631. [CrossRef] [PubMed]
4. Foster, R.G. Sleep, circadian rhythms and health. *Interface Focus* **2020**, *10*, 20190098. [CrossRef] [PubMed]
5. Duclos, C.; Beauregard, M.P.; Bottari, C.; Ouellet, M.C.; Gosselin, N. The impact of poor sleep on cognition and activities of daily living after traumatic brain injury: A review. *Aust. Occup. Ther. J.* **2015**, *62*, 2–12. [CrossRef] [PubMed]
6. Van Someren, E.J.; Cirelli, C.; Dijk, D.J.; Van Cauter, E.; Schwartz, S.; Chee, M.W. Disrupted sleep: From molecules to cognition. *J. Neurosci.* **2015**, *35*, 13889–13895. [CrossRef]
7. Mullington, J.M.; Simpson, N.S.; Meier-Ewert, H.K.; Haack, M. Sleep loss and inflammation. *Best Pract. Res. Clin. Endocrinol. Metab.* **2010**, *24*, 775–784. [CrossRef]
8. Hurtado-Alvarado, G.; Pavón, L.; Castillo-Garcia, S.A.; Hernández, M.E.; Dominguez-Salazar, E.; Velázquez-Moctezuma, J.; Gómez-González, B. Sleep loss as a factor to induce cellular and molecular inflammatory variations. *Clin. Dev. Immunol.* **2013**, *2013*, 801341. [CrossRef]
9. Lahtinen, A.; Puttonen, S.; Vanttola, P.; Viitasalo, K.; Sulkava, S.; Pervjakova, N.; Joensuu, A.; Salo, P.; Toivola, A.; Härmä, M.; et al. A distinctive DNA methylation pattern in insufficient sleep. *Sci. Rep.* **2019**, *9*, 1–9. [CrossRef]
10. He, J.; Hsuchou, H.; He, Y.; Kastin, A.J.; Wang, Y.; Pan, W. Sleep restriction impairs blood-brain barrier function. *J. Neurosci.* **2014**, *34*, 14697–14706. [CrossRef]
11. Everson, C.A.; Bergmann, B.M.; Rechtschaffen, A. Sleep deprivation in the rat: III. Total sleep deprivation. *Sleep* **1989**, *12*, 13–21. [CrossRef] [PubMed]
12. Ross, J.J. Neurological findings after prolonged sleep deprivation. *Arch. Neurol.* **1965**, *12*, 399–403. [CrossRef] [PubMed]
13. Semyachkina-Glushkovskaya, O.; Postnov, D.; Penzel, T.; Kurths, J. Sleep as a novel biomarker and a promising therapeutic target for cerebral small vessel disease: A review focusing on Alzheimer's disease and the blood-brain barrier. *Int. J. Mol. Sci.* **2020**, *21*, 6293. [CrossRef]
14. Peng, C.-K.; Buldyrev, S.V.; Havlin, S.; Simons, M.; Stanley, H.E.; Goldberger, A.L. Mosaic organization of DNA nucleotides. *Phys. Rev. E* **1994**, *49*, 1685–1689. [CrossRef]
15. Peng, C.-K.; Havlin, S.; Stanley, H.E.; Goldberger, A.L. Quantification of scaling exponents and crossover phenomena in nonstationary heartbeat time series. *Chaos* **1995**, *5*, 82–87. [CrossRef] [PubMed]
16. Stanley, H.E.; Amaral, L.A.N.; Goldberger, A.L.; Havlin, S.; Ivanov, P.C.; Peng, C.-K. Statistical physics and physiology: Monofractal and multifractal approaches. *Phys. A* **1999**, *270*, 309–324. [CrossRef]
17. Ivanova, K.; Ausloos, M. Application of the detrended fluctuation analysis (DFA) method for describing cloud breaking. *Phys. A* **1999**, *274*, 349–354. [CrossRef]
18. Heneghan, C.; McDarby, G. Establishing the relation between detrended fluctuation analysis and power spectral density analysis for stochastic processes. *Phys. Rev. E* **2000**, *62*, 6103–6110. [CrossRef] [PubMed]
19. Talkner, P.; Weber, R.O. Power spectrum and detrended fluctuation analysis: Application to daily temperatures. *Phys. Rev. E* **2000**, *62*, 150–160. [CrossRef]
20. Kantelhardt, W.; Koscielny-Bunde, E.; Rego, H.H.A.; Havlin, S.; Bunde, A. Detecting long-range correlations with detrended fluctuation analysis. *Phys. A* **2001**, *295*, 441–454. [CrossRef]
21. Frolov, N.S.; Grubov, V.V.; Maksimenko, V.A.; Lüttjohann, A.; Makarov, V.V.; Pavlov, A.N.; Sitnikova, E.; Pisarchik, A.N.; Kurths, J.; Hramov, A.E. Statistical properties and predictability of extreme epileptic events. *Sci. Rep.* **2019**, *9*, 7243. [CrossRef] [PubMed]

22. Pavlov, A.N.; Runnova, A.E.; Maksimenko, V.A.; Pavlova, O.N.; Grishina, D.S.; Hramov, A.E. Detrended fluctuation analysis of EEG patterns associated with real and imaginary arm movements. *Phys. A* **2018**, *509*, 777–782. [CrossRef]
23. Hu, K.; Ivanov, P.C.; Chen, Z.; Carpena, P.; Stanley, H.E. Effect of trends on detrended fluctuation analysis. *Phys. Rev. E* **2001**, *64*, 011114. [CrossRef] [PubMed]
24. Chen, Z.; Ivanov, P.C.; Hu, K.; Stanley, H.E. Effect of nonstationarities on detrended fluctuation analysis. *Phys. Rev. E* **2002**, *65*, 041107. [CrossRef]
25. Bryce, R.M.; Sprague, K.B. Revisiting detrended fluctuation analysis. *Sci. Rep.* **2012**, *2*, 315. [CrossRef]
26. Shao, Y.H.; Gu, G.F.; Jiang, Z.Q.; Zhou, W.X.; Sornette, D. Comparing the performance of FA, DFA and DMA using different synthetic long-range correlated time series. *Sci. Rep.* **2012**, *2*, 835. [CrossRef]
27. Pavlov, A.N.; Pavlova, O.N.; Semyachkina-Glushkovskaya, O.V.; Kurths, J. Extended detrended fluctuation analysis: Effects of nonstationarity and application to sleep data. *Eur. Phys. J. Plus* **2021**, *136*, 10. [CrossRef]
28. Kantelhardt, J.W.; Zschiegner, S.A.; Koscielny-Bunde, E.; Havlin, S.; Bunde, A.; Stanley, H.E. Multifractal detrended fluctuation analysis of nonstationary time series. *Phys. A* **2002**, *316*, 87–114. [CrossRef]
29. Castiglioni, P.; Faini, A. A fast DFA algorithm for multifractal multiscale analysis of physiological time series. *Front. Physiol.* **2019**, *10*, 115. [CrossRef]
30. Pavlov, A.N.; Abdurashitov, A.S.; Koronovskii, A.A., Jr.; Pavlova, O.N.; Semyachkina-Glushkovskaya, O.V.; Kurths, J. Detrended fluctuation analysis of cerebrovascular responses to abrupt changes in peripheral arterial pressure in rats. *Commun. Nonlinear Sci. Numer. Simul.* **2020**, *85*, 105232. [CrossRef]
31. Pavlov, A.N.; Dubrovsky, A.I.; Koronovskii, A.A., Jr.; Pavlova, O.N.; Semyachkina-Glushkovskaya, O.V.; Kurths, J. Extended detrended fluctuation analysis of electroencephalograms signals during sleep and the opening of the blood-brain barrier. *Chaos* **2020**, *30*, 073138. [CrossRef] [PubMed]
32. Pavlov, A.N.; Dubrovsky, A.I.; Koronovskii, A.A., Jr.; Pavlova, O.N.; Semyachkina-Glushkovskaya, O.V.; Kurths, J. Extended detrended fluctuation analysis of sound-induced changes in brain electrical activity. *Chaos Solitons Fractals* **2020**, *139*, 109989. [CrossRef]
33. Reynolds, A.C.; Banks, S. Total sleep deprivation, chronic sleep restriction and sleep disruption. *Prog. Brain Res.* **2010**, *185*, 91–103. [PubMed]
34. Meerlo, P.; Sgoifo, A.; Suchecki, D. Restricted and disrupted sleep: Effects on autonomic function, neuroendocrine stress systems and stress responsivity. *Sleep Med. Rev.* **2008**, *12*, 197–210. [CrossRef]
35. Evans, J.A.; Davidson, A.J. Health consequences of circadian disruption in humans and animal models. *Prog. Mol. Biol. Transl. Sci.* **2013**, *119*, 283–323.
36. Potter, G.D.; Skene, D.J.; Arendt, J.; Cade, J.E.; Grant, P.J.; Hardie, L.J. Circadian rhythm and sleep disruption: Causes, metabolic consequences, and countermeasures. *Endocr. Rev.* **2016**, *37*, 584–608. [CrossRef]
37. Medic, G.; Wille, M.; Hemels, M.E. Short- and long-term health consequences of sleep disruption. *Nat. Sci. Sleep* **2017**, *9*, 151–161. [CrossRef]
38. Hablitz, L.M.; Vinitsky, H.S.; Sun, Q.; Stæger, F.F.; Sigurdsson, B.; Mortensen, K.N.; Lilius, T.O.; Nedergaard, M. Increased glymphatic influx is correlated with high EEG delta power and low heart rate in mice under anesthesia. *Sci. Adv.* **2019**, *5*, eaav5447. [CrossRef]
39. Achariyar, T.M.; Li, B.; Peng, W.; Verghese, P.B.; Shi, Y.; McConnell, E.; Benraiss, A.; Kasper, T.; Song, W.; Takano, T.; et al. Glymphatic distribution of CSF-derived apoE into brain is isoform specific and suppressed during sleep deprivation. *Mol. Neurodegen.* **2016**, *11*, 74. [CrossRef]
40. Zhang, J.; Zhu, Y.; Zhan, G.; Fenik, P.; Panossian, L.; Wang, M.M.; Reid, S.; Lai, D.; Davis, J.G.; Baur, J.A.; et al. Extended wakefulness: Compromised metabolics in and degeneration of locus ceruleus neurons. *J. Neurosci.* **2014**, *34*, 4418–4431. [CrossRef]

Event-Related Coherence in Visual Cortex and Brain Noise: An MEG Study

Parth Chholak [1,*,†], Semen A. Kurkin [2,†], Alexander E. Hramov [2] and Alexander N. Pisarchik [1,2]

1. Centre for Biomedical Technology, Technical University of Madrid, Madrid 28223, Spain; alexander.pisarchik@ctb.upm.es
2. Centre for Technologies in Robotics and Mechatronics Components, Innopolis University, 420500 Innopolis, Russia; s.kurkin@innopolis.ru (S.A.K.); a.hramov@innopolis.ru (A.E.H.)
* Correspondence: parth.chholak@ctb.upm.es
† These authors contributed equally to this work.

Abstract: The analysis of neurophysiological data using the two most widely used open-source MATLAB toolboxes, FieldTrip and Brainstorm, validates our hypothesis about the correlation between event-related coherence in the visual cortex and neuronal noise. The analyzed data were obtained from magnetoencephalography (MEG) experiments based on visual perception of flickering stimuli, in which fifteen subjects effectively participated. Before coherence and brain noise calculations, MEG data were first transformed from recorded channel data to brain source waveforms by solving the inverse problem. The inverse solution was obtained for a 2D cortical shape in Brainstorm and a 3D volume in FieldTrip. We found that stronger brain entrainment to the visual stimuli concurred with higher brain noise in both studies.

Keywords: MEG; FieldTrip; Brainstorm; source reconstruction; flickering; cognitive neuroscience; visual perception

1. Introduction

The human brain is a complex network consisting of approximately 86 billion neurons [1] subdivided into oscillatory clusters that fire co-dependently/independently to manifest our consciousness as we know it. These clusters correspond to regions of the brain specialized in processing certain types of information and are connected to other specialized regions in complex networks.

Brain connectivity is studied in three forms: functional, structural, and effective [2–5]. Structural connectivity identifies anatomical neural networks that show possible pathways for neural communication [6,7]. Functional connectivity finds active brain regions that have a correlated frequency, phase, and/or amplitude [8]. Effective connectivity utilizes the functional connectivity information and additionally determines the direction of the dynamic information flow [9,10]. Effective and functional connectivity can be measured in the frequency domain (e.g., coherence [11]) and in the time domain (e.g., Granger causality [5] or artificial neuronal network-based functional connectivity [12]).

Inter-neuronal communication is realized by one of 50+ neurotransmitters that can be either excitatory (e.g., dopamine) or inhibitory (e.g., gamma-Aminobutyric acid (GABA)) [13]. Voltage-gated ion channels on the cell membranes of neurons generate action potentials and periodic membrane potential activity that synchronizes neighboring neurons [14,15]. These neighboring neurons may, in turn, affect other remotely located neurons, creating a network of connectivity. Coherence-based neuronal communications are driven by the dynamics of neurotransmitters such as amino acid glutamate and GABA.

Only when the synchronous neuronal population is large enough, the produced electrical activity and the concomitant magnetic activity is strong enough to be detected outside the skull using methods such as electroencephalography (EEG) and magnetoencephalography (MEG), respectively [16]. MEG measures the ionic currents inside the neuron (primary currents), whereas EEG measures the return or volume currents outside the neuron (secondary currents).

Coherence is commonly used to quantify neuronal synchronicity between spatially separated EEG electrodes or MEG coils [17]. It is essentially an estimate of the consistency of the relative amplitude and phase between two signals within a given frequency band. There is a linear mathematical method resulting in a symmetric matrix, lacking any directional information. Identical signals produce a coherence magnitude of 1, whereas the coherence magnitude approaches 0 as the dissimilarity between the considered signals increases.

In the last two decades, significant progress has been achieved in the development of new computational algorithms that enable connectivity calculations directly between the different regions of the brain (source space) [5] instead of electrodes or coils (channel space). The source space analysis provides better anatomical localization [18] and enables inter-subject or group analysis as the brain activity now can be projected onto a more standardized space.

In 2004, Hoechstetter et al. [19] introduced a new method to study source coherence in the brain. Discrete multiple source models were created using brain electrical source analysis, and the source activity was transformed into time–frequency space. Finally, magnitude-squared coherence was evaluated to reveal coupled brain sources. The application of inverse solutions to estimate brain activity in the source space from the channel space removes current leakage among adjacent channels. This averts localization errors that are fundamental to coherence analysis in the channel space [19].

Coherence has henceforth been used in many brain connectivity studies on patients and healthy subjects, including but not limited to studies on working memory [20], brain lesions [21], hemiparesis [22], resting-state networks [23], schizophrenia [24–26], favorable responses to panic medications [27], and motor imagery [28,29].

Owing to the diversity of human brains, we observe various forms of coherent neuronal activity over different subjects in response to the same flickering stimulus. For example, the presentation of flickering visual stimuli instils coherent responses in the visual cortex at the flicker frequency and its harmonics with varying coherent neuronal network sizes among the subjects [30,31].

Another signal processing technique used to measure synchronization in EEG and MEG is phase synchronization, a measure of how stable the phase difference is over the considered time duration. Phase synchronization requires considered signals to be phase-locked with zero or any finite phase difference, regardless of their respective amplitudes. This is in contrast to the coherence measurement in which phase and amplitude are intertwined for its estimation [32].

Noise, as known [33], can cause desynchronization in a neuronal network. Each participating neuron and interconnecting synapses add to the inherent brain noise when a stimulus is presented to the subject. Therefore, one could argue that a larger neuronal network would carry a higher brain noise and, consequently, lower average coherence. On the other hand, larger active neuronal oscillations in response to the stimulus are likely to have stronger average coherent activity and would also entail higher brain noise. Thus, the relation between the observed coherence and the level of inherent brain noise remains unclear and is the central problem explored in this paper (for comprehensive theoretical descriptions, see [34,35]).

Recently, an approach to estimate inherent brain noise based on phase synchronization was proposed [30]. The method is based on the experiments with flickering images and simultaneous recording of magnetoencephalographic (MEG) data. This paper utilizes the same methodology to measure brain noise using the same experimental paradigm and reveal its correlation with the induced coherence or source power in the visual cortex. We deal with the two most popular open-source MATLAB toolboxes for MEG data analysis, namely FieldTrip [36] and Brainstorm [37], to perform two independent analyses that are more suitable to each software.

2. Materials and Methods

We carried out MEG experiments based on the flickering paradigm with 17 conditionally healthy subjects (age: 17–64 years; 10 males) with normal or corrected-to-normal visual acuity. Two subjects were later discarded. Frequency tags at the stimulus frequency and its harmonics were absent in subject "sub08", perhaps due to a lack of focus on the experiment. Meanwhile, for subject "sub11", the ECG activity was not recorded during the experiment due to a technical error, and therefore the signal-to-noise ratio of the subsequently cleaned data was too low to allow correct data analysis. All subjects provided written informed consent before the experiment commencement. The experiments were performed as per the Declaration of Helsinki and approved by the Ethics Committee of the Technical University of Madrid.

2.1. MEG Acquisition

MEG recordings were performed with an Elekta-Neuromag system with 306 channels that was housed in a magnetically shielded room at the Centro de Tecnología Biomédica, Universidad Politécnica de Madrid. The head position was continuously tracked with head position indicator (HPI) coils and co-registered in the device and head coordinate system with three fiducial points (nasion, left, and right preauricular points) and around 300 scalp surface points digitized by a Polhemus Fastrak system. A vertical electrooculogram (EOG) and electrocardiogram (ECG) were placed to capture eye blinks and cardiac activity, respectively. The data were sampled at 1000 Hz.

The experiments for all 17 subjects lasted 4 days. Along with MEG recordings of the subjects, the MEG data were also collected daily in an empty room. All data were passed through an online anti-alias bandpass (0.1–330)-Hz filter. MaxFilter software was used for the temporal signal-space separation (tSSS) to reduce magnetic interference and perform head movement compensation. A 56-ms delay between event triggers and the actual stimulus was measured separately using a photodiode.

2.2. Flickering Stimulation

A grey square image with varying greyness levels on a grey background (brightness: 127 in 8-bit format) was projected onto a translucent screen positioned 150 cm away from the subjects with a 60-Hz frame rate. The pixels' brightness was modulated by a harmonic signal with frequency $f_m = 6.67$ Hz (60/9) and a 50% amplitude, i.e., between black (0) and grey (127). This particular frequency was chosen because it produces the most pronounced spectral response in the visual cortex [30].

2.3. Experimental Protocol

The participants were informed about the experimental protocol beforehand in addition to the corresponding textual directives on the screen throughout the experiment. The experiment started with the presentation of a static (non-flickering) square image with a red dot at the center, on which the participant had to concentrate their gaze for 120 s. The recorded brain activity was used as a reference signal or background. After a short rest, the square image started flickering and was presented 2–5 times (depending on the subject) for 120 s, interrupted by a 30-s resting period between each presentation. The flickering stimulus was presented at least 3 times to all subjects except for subject "sub10". The starting times of the background and flickering recordings were marked with event triggers using a parallel-port setup.

2.4. Analysis Pipelines in FieldTrip and Brainstorm

Next, we will discuss the common steps of the MEG data analysis and related implementation details in both FieldTrip and Brainstorm software. We will focus on the analysis of our experimental data, which, of course, does not cover the full functionality of the two toolboxes.

2.4.1. Reading and Segmenting Data

We start our analysis by reading the MEG data stored in the FIFF format and segmenting them into trials according to experimental conditions. It is common to segment the data after decoding trigger sequences in a raw data file. However, in this work, we make use of additional functions to import events from mat-files in both analyses because there are slight differences in the experimental protocol for some subjects, and this approach is more time-efficient than if-else conditions specifying the subjective exceptions in the batch-processing scripts.

After extracting 120-s epochs for both experimental conditions, including the background activity trial called "B-trial" and event-related trials called "F-trials", we split every 120-s trial into 4-s (for FieldTrip; see explanation below) or 3-s (for Brainstorm) sub-trials.

2.4.2. Artifact Removal and Loading Data

Accurate brain source analysis requires the correct integration of MEG data with structural magnetic resonance imaging (MRI) scans. Both software programs align all data by defining a subject coordinate system using three fiducial points, namely the nasion and left- and right-auricular points. Moreover, we complemented the alignment based on only three points by an automatic refinement procedure utilizing additional points on the scalp, marked using a 3D digitizer (Polhemus in the considered experiment).

In FieldTrip, we used the "Colin27" head averaged template MRI [38] and adjusted it to the subject's head shape recorded by the Polhemus device. In Brainstorm, default anatomy was warped to fit the scalp shape of every subject with a 2% fit tolerance using digitized head points from the Polhemus device. After an automatic refinement of head points, the 50-Hz electrical power grid frequency and its harmonics were filtered using notch filters. The 56-ms trigger delay was corrected in the recordings. The recorded electrooculogram (EOG) and electrocardiogram (ECG) signals were used to automatically detect instances of eye blinks and cardiac activity in order to apply signal-space projection (SSP) methods to alleviate the respective artifacts.

Well-defined artifacts such as eye blinks, cardiac activity, muscle contractions, and MEG SQUID jumps were detected semi-automatically using FieldTrip/Brainstorm functions or manual screening. Once artifacts were identified, depending on the artifact intensity, we either discarded trials that contained an artifact or applied linear projection to remove them.

2.4.3. Source Reconstruction

The first step in localizing sources is the construction of a forward model and lead field matrix. The forward model allows one to calculate an estimate of the field measured by the MEG sensors for a given current distribution in the brain and is typically constructed for each subject. The lead fields or the solution to the forward problem are evaluated using various algorithms, such as a single sphere [39], overlapping spheres [40], a spherical harmonics approximation of realistic geometries [41], and boundary element methods [42].

The forward solution was computed in Brainstorm analysis using the overlapping spheres method, which is the default. The number of cortical sources was kept at 15,000 as recommended [37]. On the other hand, in FieldTrip, we applied a semi-realistic head model developed by Nolte [41] called a single-shell model, which is based on the correction of the lead field for a spherical volume conductor by a superposition of basic functions, gradients of harmonic functions constructed from spherical harmonics. We thus discretized the head volume with a grid with a 0.7-cm resolution and obtained a source space consisting of

9025 voxels. The lead field matrix was calculated using each grid point [41]. Thus, in Brainstorm, a cortical surface model was used, and in FieldTrip, a volumetric one.

The next step is calculating the inverse solution to estimate the location and strength of neuronal activity, which can be computed via multiple options, including dipole fitting based on nonlinear optimization [43], minimum variance beamformers in time and frequency domains [44–46], and linear estimation of distributed source models [47,48]. In both software analyses, we used standardized low-resolution brain electromagnetic tomography (sLORETA) [49].

The sLORETA family of solutions was validated against numerous imaging modalities [50–52] and simulations [53,54]. sLORETA uses standardized current density images to calculate intra-cerebral generators. Although the image was blurred, sLORETA was found [55] to have the exact zero-error localization when reconstructing single sources in all noise-free simulations, i.e., the maximum of the current density power estimate coincided with the exact dipole location [48]. Meanwhile, in all simulations with noise, sLORETA had the lowest localization errors when compared with the minimum norm solution.

Note that when working with multiple sensor types to form a joint source model, the empirical noise covariance is used to compute the weights of each sensor in the overall model. For this purpose, noise covariance matrices are typically computed from empty-room recordings that capture instrumental and environmental noise in the absence of subjects.

2.4.4. Event-Related Coherence

Stimulus-induced coherence in the brain was used to estimate activated brain network size and characterize its activation strength. The taken approach was different for each software program. The previous study with the same stimulus [30] found frequency tags at the flickering frequency (6.67 Hz) and its harmonics. The study also revealed that the frequency tags were more pronounced at the second harmonic (13.33 Hz) than at the first harmonic. Therefore, we need to find an index characterizing the spectral power of brain response at the second harmonic. Since the analysis methods and obtained source models are significantly different, we would require appropriate and independent indices for both software programs to estimate event-related coherence (ERC).

In FieldTrip, we first applied a fourth-order Butterworth 13–14 Hz band-pass filter. The band-pass frequency was determined by the frequency of interest, which in our case was equal to 13.33 Hz (the second harmonic of the flicker frequency). Then, we redefined the 3-s length of every trial within the [0.5, 3.5]-s interval to reduce edge effects due to filtering. Moreover, in this step, we calculated the covariance matrix, necessary when using the sLORETA method. After performing reconstruction of the sources separately for all 3-s B- and F-sub-trials using the sLORETA method, we obtained the power distribution of the activity of the brain sources on the 3D grid with 9025 voxels for every sub-trial.

In the next step, we averaged the resulting source power distributions and obtained a distribution each for B- and F-condition (P_B and P_F). Such an approach made it possible to reduce the influence of instrumental and brain noise on the results of source reconstruction and, thus, increase the prominence of the event-related pattern of neural activity, compared to source reconstruction based on a single original 120-s trial. After that, we calculated the normalized difference of the source power distributions for F- and B-conditions (so-called baseline correction): $D = (P_F - P_B)/P_B$. This procedure was needed to isolate the event-related pattern of source activity. The above steps were repeated for three 120-s F-trials (all subjects had at least three F-trials except "sub10"), and the average of the obtained three differences D was calculated. Finally, the distribution of the averaged difference was interpolated on the used MRI image.

In Brainstorm, we started by calculating magnitude-squared coherence between the time series of each of the 15,000 brain sources and the reference sinusoidal signal at frequency $2f_m$ (13.33 Hz), i.e., the second harmonic of the flicker frequency, for both the F-trials (C^F) and the B-trial (C^B). To evaluate the event-related coherence in the brain, we calculated differences between the coherence values of the F-trials and B-trials for cortical sources lying in visual areas V_1 and V_2 as per the Brodmann atlas, $ERC = C_{vis}^F - C_{vis}^B$, and averaged it for each subject.

2.5. Brain Noise Estimation

The proposed brain noise estimation method is based on phase synchronization, which implies a measurement of a phase difference between the brain's response in the visual cortex and the reference signal at the second harmonic frequency ($2f_m = 13.33$ Hz). First, to obtain the visual response, we averaged the source activity waveforms from the V_1 and V_2 subregions of the Brodmann atlas for each of the F-trials of a subject. We then bandpass-filtered this average visual response in the 13–14 Hz frequency band. To estimate brain noise, we calculated the phase difference time series between visual response time series and the second harmonic of the flicker sinusoidal signal, as [30,56]:

$$\Phi = (t_n^V - t_n^m)2f_m, \quad (1)$$

where t_n^V and t_n^m are the times of nth maxima of the visual response time series and the second harmonic of the flicker signal, respectively. Intermittent frequency-locking was observed, superposed with random fluctuations due to phase noise [33]. We also obtained unimodal probability distributions of these phase differences Φ to characterize the phase-noise-induced random fluctuations in phase. Kurtosis, a measure of the sharpness of a unimodal distribution, would be lower for a broader and noisier phase fluctuation distribution, and vice versa. Therefore, from the probability distribution of these random phase fluctuations, we estimated brain noise as the inverse distribution's kurtosis. This method was comprehensively described in the previous paper [30].

3. Results

Based on the obtained normalized distributions of the source power, we calculated for each subject the average power of source activity in the visual cortex, D_{avg}, in FieldTrip. It should be noted that we determined the visual cortex using the automated anatomical labeling (AAL) brain atlas [57] in FieldTrip. The average spectral power D_{avg} in the visual cortex was plotted in Figure 1 against estimated brain noise to phase synchronization (in units of inverse kurtosis) for every subject.

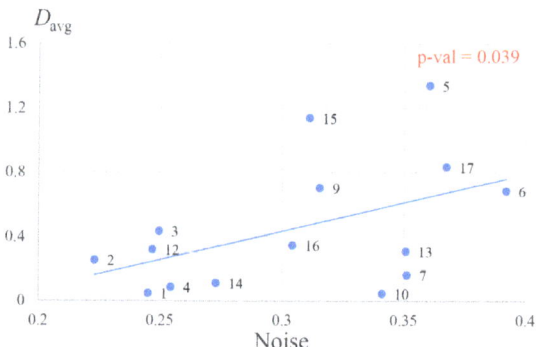

Figure 1. The average power of source activity in the brain volume corresponding to visual cortex versus brain noise for all subjects (numbers denote the subjects). The line is a linear approximation fit ($p = 0.039$, $R^2 = 0.309$).

One can see a linear correlation (with p-value equal to 0.039 and an R^2-value of 0.309) of D_{avg} and noise level, although the scatter is significant: $D_{avg}^{min} = 0.04$, $D_{avg}^{max} = 1.34$; $Noise^{min} = 0.22$, $Noise^{max} = 0.39$.

Figure 2 shows typical distributions of normalized source power D predominantly activated within the visual cortex for subjects with low (subject 2) and high (subject 6) brain noise. Subject 6 is characterized by more pronounced high-amplitude activity spanning a larger volume in the visual cortex than subject 2.

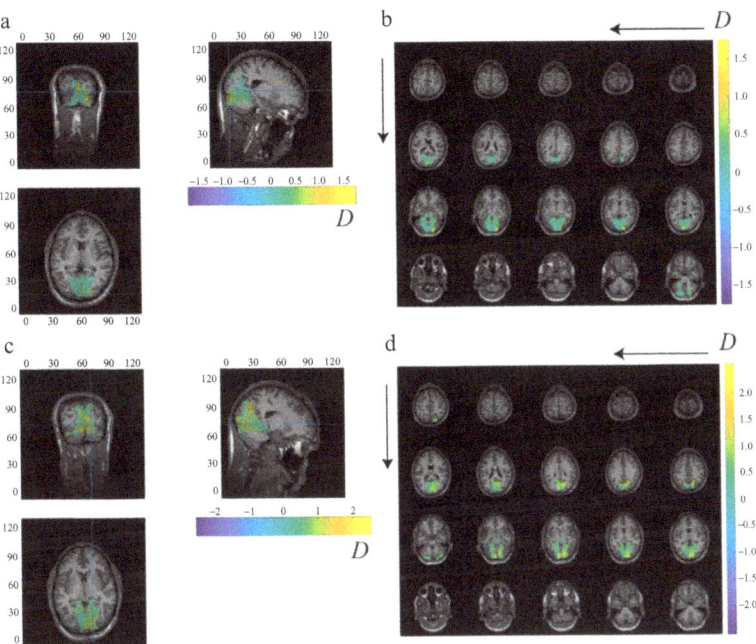

Figure 2. Distributions of normalized source power D within the visual cortex plotted superimposed on anatomical MRI in orthogonal cut view (**a**,**c**) and slice mode (**b**,**d**) for subject 2 (**a**,**b**) and subject 6 (**c**,**d**). The blue crosses in (**a**,**c**) define the cutting planes. The arrows in (**b**,**d**) indicate the direction of movement along the slices.

We will show now the results of the alternate analysis pipeline in Brainstorm. The values of average event-related coherence over visual areas V1 and V2 were compared with the same estimated brain noise as used in Figure 1 for all subjects. A linear relation was established with a p value of 0.048 and an R^2-value of 0.267, as seen in Figure 3. The distributions of average event-related coherence over the cortex for typical subjects with low and high noise levels are shown in Figure 4 as per the cortical analysis in Brainstorm.

The methodology to calculate the normalized difference of power on a 3D volume in FieldTrip was adapted to fit the 2D source model generated in Brainstorm to have a closer comparison. Figure 5 shows the corresponding linear regression model with a p-value of 0.209 and an R^2-value of 0.118 ($D_{avg}^{min} = 0.08$, $D_{avg}^{max} = 2.18$; $Noise^{min} = 0.22$, $Noise^{max} = 0.39$). Although the model fails to capture any significant relation, the relative positions of the subjects in the power–noise state-space of Figure 5 are quite similar to those which we observe in Figure 1.

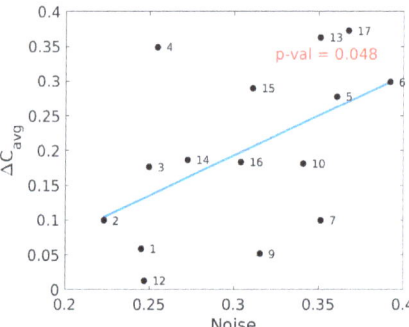

Figure 3. Average event-related coherence in the visual cortex versus estimated brain noise to phase synchronization. The straight line is a linear regression fit of the data ($p = 0.048$, $R^2 = 0.267$).

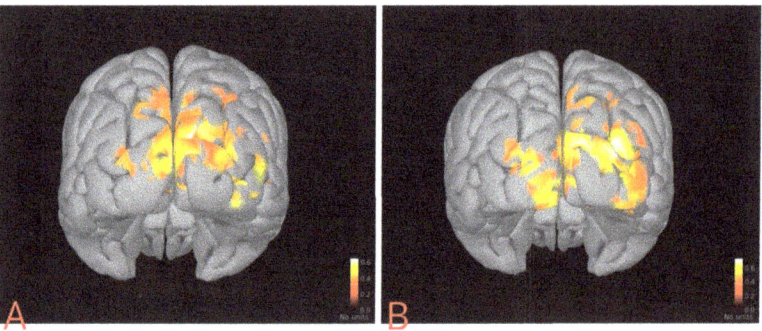

Figure 4. Typical cortical distributions of event-related coherence for (**A**) Subject 2 (low noise) and (**B**) Subject 6 (high noise). The brain activation is more intensive in the latter case.

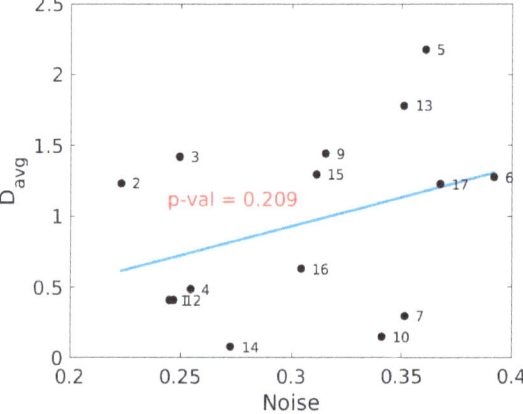

Figure 5. The average power of source activity in the visual cortex versus brain noise for all subjects (numbers denote the subjects). The line is a linear approximation fit ($p = 0.209$, $R^2 = 0.118$).

4. Discussion

We found a linear relation, with a positive slope between the average power of source activity in the visual cortex and brain noise. The results show that the subjects with more powerful visual cortex activity demonstrate more substantial brain noise. This relationship can be explained as follows. The higher the power of the reconstructed sources, the more neurons are involved in realizing cognitive activity. In a larger network of neurons, the number of synapses would also be higher, and both the synapses and the neurons would feed the phase-destabilizing noise into the system [30].

The two independent methods essentially lead to the calculation of the difference in spectral activity inside the subject's brain, corresponding to the second harmonic of the stimulus frequency when the subject is observing a flickering image, as opposed to when the subject is gazing at a stationary stimulus. Averaging them over the respective regions of interest led to very similar trends between average event-related coherence or frequency-filtered signal power and brain noise using either software program (Figures 1 and 3). One can see in Figures 2 and 4 that the subject with higher brain noise ("sub06") has a more extensive and intensely activated neuronal network, coherent with the stimulus, as distinct from the subject with lower brain noise ("sub02").

As we have already mentioned above, we set out to adapt the prescribed analysis pipelines of both FieldTrip and Brainstorm to our study. The two software programs gave congruent results following their independent analysis strategies. However, it should not be a surprise that if we try mixing the two analysis pipelines midway, the results will likely deteriorate. Figure 5 shows the result of such mindless mixing of the two methods. Even though the order of subjects' frequency-filtered signal powers remained conserved from Figure 1, the linear relation was lost.

Since we calculate brain noise from the phase fluctuation time series and the corresponding probability distribution, which in turn depends upon the signal-to-noise ratio (SNR) of the source waveforms in the visual cortex to be properly calculated, it can turn into a circular problem where, for very high brain noises, the SNR would be too low to correctly determine the phase fluctuations, which would make the calculation of brain noise impossible. This was the case with subjects "sub8" and "sub11". For these subjects, we did not see any frequency tags in the power spectrum during the flickering cube presentation (signal) and also in the power spectrum for the stationary cube presentation (noise). Thus, they had to be removed from the study. The subjects who showed frequency tags in the power spectrum also had clear bandpass-filtered waveforms in the 13–14 Hz frequency band used to calculate phase difference fluctuations.

We have to emphasize that all codes of our analysis and MEG data used for this study were made publicly available during the review period. The developed methods, along with the prescribed codes on the software documentations adapted to a generic MEG study starting with only a FIFF file, will be accessible to newcomers in the field.

5. Conclusions

Visual flickering experiments were carried out successfully with fifteen healthy subjects, and their brain responses were recorded using MEG. The two most popular open-source software programs, FieldTrip and Brainstorm, were used to analyze brain source activity. We calculated the event-related coherence of the brain response with the flickering visual signal. Using a recently proposed brain noise estimation method, we computed the relation between the coherent brain network in the visual cortex and corresponding brain noise. The results obtained by the two software programs demonstrated fair agreement. The analyses performed by both MATLAB toolboxes evidenced that more extensive brain activity is accompanied by more substantial brain noise.

Author Contributions: Conceptualization, P.C. and S.A.K.; methodology, A.N.P. and A.E.H.; software, P.C. and S.A.K.; validation, A.N.P. and A.E.H.; formal analysis, all authors; investigation, A.N.P. and A.E.H.; resources, A.N.P.; data curation, P.C.; writing—original draft preparation, P.C. and S.A.K.; writing—review and editing, all authors; visualization, P.C. and S.A.K.; supervision, A.N.P. and A.E.H.; project administration, A.N.P. and A.E.H.; funding acquisition, A.N.P. and A.E.H. All authors have read and agreed to the published version of the manuscript.

Funding: This research was funded by the Spanish Ministry of Economy and Competitiveness SAF2016-80240 in the part of the experimental study and Russian Science Foundation No. 19-12-00050 in the part of the data processing. S.A.K. is supported by the President Program (MD-1921.2020.9) in the part of the source reconstruction.

Institutional Review Board Statement: The study was conducted according to the guidelines of the Declaration of Helsinki, and approved by the Ethics Committee of the Technical University of Madrid (protocol No. 19, 3 February 2020).

Informed Consent Statement: Informed consent was obtained from all subjects involved in the study.

Data Availability Statement: The MEG data have been uploaded at https://zenodo.org/record/4408648#.X-72UdYo-Cc. Both FieldTrip and Brainstorm codes have been uploaded as a single release at https://zenodo.org/record/4408756#.X-77itYo-Cc.

Acknowledgments: The authors graciously acknowledge the subjects for their contribution to this research.

Conflicts of Interest: The authors declare no conflict of interest. The funders had no role in the design of the study; in the collection, analyses, or interpretation of data; in the writing of the manuscript, or in the decision to publish the results.

References

1. Azevedo, F.A.C.; Carvalho, L.R.B.; Grinberg, L.T.; Farfel, J.M.; Ferretti, R.E.L.; Leite, R.E.P.; Jacob Filho, W.; Lent, R.; Herculano-Houzel, S. Equal numbers of neuronal and nonneuronal cells make the human brain an isometrically scaled-up primate brain. *J. Comp. Neurol.* **2009**, *513*, 532–541. [CrossRef] [PubMed]
2. Friston, K.J.; Frith, C.D.; Liddle, P.F.; Frackowiak, R.S. Functional connectivity: The principal-component analysis of large (PET) data sets. *J. Cereb. Blood Flow Metab.* **1993**, *13*, 5–14. [CrossRef] [PubMed]
3. Greenblatt, R.E.; Pflieger, M.E.; Ossadtchi, A.E. Connectivity measures applied to human brain electrophysiological data. *J. Neurosci. Methods* **2012**, *207*, 1–16. [CrossRef] [PubMed]
4. Sakkalis, V. Review of advanced techniques for the estimation of brain connectivity measured with EEG/MEG. *Comput. Biol. Med.* **2011**, *41*, 1110–1117. [CrossRef] [PubMed]
5. Hramov, A.E.; Frolov, N.S.; Maksimenko, V.A.; Kurkin, S.A.; Kazantsev, V.B.; Pisarchik, A.N. Functional networks of the brain: From connectivity restoration to dynamic integration. *Phys. Uspekhi* **2020**, *63*. [CrossRef]
6. Le Bihan, D.; Mangin, J.F.; Poupon, C.; Clark, C.A.; Pappata, S.; Molko, N.; Chabriat, H. Diffusion tensor imaging: Concepts and applications. *J. Magn. Reson. Imaging JMRI* **2001**, *13*, 534–546. [CrossRef]
7. Wedeen, V.J.; Wang, R.P.; Schmahmann, J.D.; Benner, T.; Tseng, W.Y.I.; Dai, G.; Pandya, D.N.; Hagmann, P.; D'Arceuil, H.; de Crespigny, A.J. Diffusion spectrum magnetic resonance imaging (DSI) tractography of crossing fibers. *NeuroImage* **2008**, *41*, 1267–1277. [CrossRef]
8. Towle, V.L.; Hunter, J.D.; Edgar, J.C.; Chkhenkeli, S.A.; Castelle, M.C.; Frim, D.M.; Kohrman, M.; Hecox, K.E. Frequency Domain Analysis of Human Subdural Recordings. *J. Clin. Neurophysiol.* **2007**, *24*, 205–213. [CrossRef]
9. Cabral, J.; Kringelbach, M.L.; Deco, G. Exploring the network dynamics underlying brain activity during rest. *Prog. Neurobiol.* **2014**, *114*, 102–131. [CrossRef]
10. Horwitz, B. The elusive concept of brain connectivity. *NeuroImage* **2003**, *19*, 466–470. [CrossRef]
11. Bowyer, S.M. Coherence a measure of the brain networks: Past and present. *Neuropsychiatr. Electrophysiol.* **2016**, *2*, 1. [CrossRef]
12. Frolov, N.; Maksimenko, V.; Lüttjohann, A.; Koronovskii, A.; Hramov, A. Feed-forward artificial neural network provides data-driven inference of functional connectivity. *Chaos Interdiscip. J. Nonlinear Sci.* **2019**, *29*, 091101. [CrossRef] [PubMed]
13. Chana, G.; Bousman, C.A.; Money, T.T.; Gibbons, A.; Gillett, P.; Dean, B.; Everall, I.P. Biomarker investigations related to pathophysiological pathways in schizophrenia and psychosis. *Front. Cell. Neurosci.* **2013**, *7*. [CrossRef] [PubMed]
14. Llinás, R.R. The intrinsic electrophysiological properties of mammalian neurons: insights into central nervous system function. *Science* **1988**, *242*, 1654–1664. [CrossRef]
15. Llinás, R.R.; Grace, A.A.; Yarom, Y. In vitro neurons in mammalian cortical layer 4 exhibit intrinsic oscillatory activity in the 10- to 50-Hz frequency range. *Proc. Natl. Acad. Sci. USA* **1991**, *88*, 897–901. [CrossRef]
16. Hämäläinen, M.; Hari, R.; Ilmoniemi, R.J.; Knuutila, J.; Lounasmaa, O.V. Magnetoencephalography—Theory, instrumentation, and applications to noninvasive studies of the working human brain. *Am. Phys. Soc.* **1993**, *65*. [CrossRef]
17. French, C.C.; Beaumont, J.G. A critical review of EEG coherence studies of hemisphere function. *Int. J. Psychophysiol. Off. J. Int. Organ. Psychophysiol.* **1984**, *1*, 241–254. [CrossRef]
18. Kurkin, S.; Hramov, A.; Chholak, P.; Pisarchik, A. Localizing oscillatory sources in a brain by MEG data during cognitive activity. In Proceedings of the 2020 4th International Conference on Computational Intelligence and Networks (CINE), Kolkata, India, 27–29 February 2020; pp. 1–4.
19. Hoechstetter, K.; Bornfleth, H.; Weckesser, D.; Ille, N.; Berg, P.; Scherg, M. BESA source coherence: A new method to study cortical oscillatory coupling. *Brain Topogr.* **2004**, *16*, 233–238. [CrossRef]

20. Gross, J.; Schmitz, F.; Schnitzler, I.; Kessler, K.; Shapiro, K.; Hommel, B.; Schnitzler, A. Modulation of long-range neural synchrony reflects temporal limitations of visual attention in humans. *Proc. Natl. Acad. Sci. USA* **2004**, *101*, 13050–13055. [CrossRef]
21. Guggisberg, A.G.; Honma, S.M.; Findlay, A.M.; Dalal, S.S.; Kirsch, H.E.; Berger, M.S.; Nagarajan, S.S. Mapping functional connectivity in patients with brain lesions. *Ann. Neurol.* **2008**, *63*, 193–203. [CrossRef]
22. Belardinelli, P.; Ciancetta, L.; Staudt, M.; Pizzella, V.; Londei, A.; Birbaumer, N.; Romani, G.L.; Braun, C. Cerebro-muscular and cerebro-cerebral coherence in patients with pre- and perinatally acquired unilateral brain lesions. *NeuroImage* **2007**, *37*, 1301–1314. [CrossRef] [PubMed]
23. de Pasquale, F.; Della Penna, S.; Snyder, A.Z.; Lewis, C.; Mantini, D.; Marzetti, L.; Belardinelli, P.; Ciancetta, L.; Pizzella, V.; Romani, G.L.; et al. Temporal dynamics of spontaneous MEG activity in brain networks. *Proc. Natl. Acad. Sci. USA* **2010**, *107*, 6040–6045. [CrossRef] [PubMed]
24. Hinkley, L.B.N.; Vinogradov, S.; Guggisberg, A.G.; Fisher, M.; Findlay, A.M.; Nagarajan, S.S. Clinical symptoms and alpha band resting-state functional connectivity imaging in patients with schizophrenia: implications for novel approaches to treatment. *Biol. Psychiatry* **2011**, *70*, 1134–1142. [CrossRef] [PubMed]
25. Kim, J.S.; Shin, K.S.; Jung, W.H.; Kim, S.N.; Kwon, J.S.; Chung, C.K. Power spectral aspects of the default mode network in schizophrenia: An MEG study. *BMC Neurosci.* **2014**, *15*, 104. [CrossRef]
26. Bowyer, S.M.; Gjini, K.; Zhu, X.; Kim, L.; Moran, J.E.; Rizvi, S.U.; Gumenyuk, N.T.; Tepley, N.; Boutros, N.N. Potential Biomarkers of Schizophrenia from MEG Resting-State Functional Connectivity Networks: Preliminary Data. *J. Behav. Brain Sci.* **2015**. *5*, 1. [CrossRef]
27. Boutros, N.N.; Galloway, M.P.; Ghosh, S.; Gjini, K.; Bowyer, S.M. Abnormal coherence imaging in panic disorder: A magnetoencephalography investigation. *Neuroreport* **2013**, *24*, 487–491. [CrossRef]
28. Chholak, P.; Pisarchik, A.N.; Kurkin, S.A.; Maksimenko, V.A.; Hramov, A.E. Neuronal pathway and signal modulation for motor communication. *Cybern. Phys.* **2019**, 106–113. [CrossRef]
29. Chholak, P.; Niso, G.; Maksimenko, V.A.; Kurkin, S.A.; Frolov, N.S.; Pitsik, E.N.; Hramov, A.E.; Pisarchik, A.N. Visual and kinesthetic modes affect motor imagery classification in untrained subjects. *Sci. Rep.* **2019**, *9*, 1–12.
30. Pisarchik, A.N.; Chholak, P.; Hramov, A.E. Brain noise estimation from MEG response to flickering visual stimulation. *Chaos Solitons Fractals X* **2019**, 100005. [CrossRef]
31. Chholak, P.; Maksimenko, V.A.; Hramov, A.E.; Pisarchik, A.N. Voluntary and involuntary attention in bistable visual perception: A MEG study. *Front. Hum. Neurosci.* **2020**, *14*, 555.
32. Uhlhaas, P.J.; Roux, F.; Rodriguez, E.; Rotarska-Jagiela, A.; Singer, W. Neural synchrony and the development of cortical networks. *Trends Cogn. Sci.* **2010**, *14*, 72–80. [CrossRef] [PubMed]
33. Boccaletti, S.; Pisarchik, A.; Genio, C.; Amann, A. *Synchronization: From Coupled Systems to Complex Networks*; Cambridge University Press: Cambridge, UK, 2018. [CrossRef]
34. Chholak, P.; Hramov, A.E.; Pisarchik, A.N. An advanced perception model combining brain noise and adaptation. *Nonlinear Dyn.* **2020**, *100*, 3695–3709. [CrossRef]
35. An, P.; Va, M.; Av, A.; Ns, F.; Vv, M.; Mo, Z.; Ae, R.; Ae, H. Coherent resonance in the distributed cortical network during sensory information processing. *Sci. Rep.* **2019**, *9*, 18325–18325. [CrossRef]
36. Oostenveld, R.; Fries, P.; Maris, E.; Schoffelen, J.M. FieldTrip: Open Source Software for Advanced Analysis of MEG, EEG, and Invasive Electrophysiological Data. *Comput. Intell. Neurosci.* **2010**, *2011*, e156869. ISSN 1687-5265. [CrossRef]
37. Tadel, F.; Baillet, S.; Mosher, J.C.; Pantazis, D.; Leahy, R.M. Brainstorm: A user-friendly application for MEG/EEG analysis. *Comput. Intell. Neurosci.* **2011**, *2011*, 879716. [CrossRef]
38. Holmes, C.J.; Hoge, R.; Collins, L.; Woods, R.; Toga, A.W.; Evans, A.C. Enhancement of MR images using registration for signal averaging. *J. Comput. Assist. Tomogr.* **1998**, *22*, 324–333. [CrossRef]
39. Cuffin, B.N.; Cohen, D. Magnetic fields of a dipole in special volume conductor shapes. *IEEE Trans. Bio-Med. Eng.* **1977**, *24*, 372–381. [CrossRef]
40. Huang, M.X.; Mosher, J.C.; Leahy, R.M. A sensor-weighted overlapping-sphere head model and exhaustive head model comparison for MEG. *Phys. Med. Biol.* **1999**, *44*, 423–440. [CrossRef]
41. Nolte, G. The magnetic lead field theorem in the quasi-static approximation and its use for magnetoencephalography forward calculation in realistic volume conductors. *Phys. Med. Biol.* **2003**, *48*, 3637–3652. [CrossRef]
42. Darvas, F.; Pantazis, D.; Kucukaltun-Yildirim, E.; Leahy, R.M. Mapping human brain function with MEG and EEG: methods and validation. *NeuroImage* **2004**, *23*, S289–299. [CrossRef]
43. Scherg, M. Fundamentals of dipole source analysis. *Adv. Audiol.* **1990**, *6*, 40–69.
44. Van Veen, B.D.; van Drongelen, W.; Yuchtman, M.; Suzuki, A. Localization of brain electrical activity via linearly constrained minimum variance spatial filtering. *IEEE Trans. Bio-Med. Eng.* **1997**, *44*, 867–880. [CrossRef] [PubMed]
45. Gross, J.; Kujala, J.; Hämäläinen, M.; Timmermann, L.; Schnitzler, A.; Salmelin, R. Dynamic imaging of coherent sources: Studying neural interactions in the human brain. *Proc. Natl. Acad. Sci. USA* **2001**, *98*, 694–699. [CrossRef] [PubMed]
46. Kurkin, S.; Chholak, P.; Pisarchik, A.; Hramov, A. Analysis of the features of brain neuronal sources during imagery motor activity: MEG study. In Proceedings of the 2020 4th Scientific School on Dynamics of Complex Networks and their Application in Intellectual Robotics (DCNAIR), Innopolis, Russia, 7–9 september 2020; pp. 154–157.

47. Hämäläinen, M.S.; Ilmoniemi, R.J. Interpreting magnetic fields of the brain: Minimum norm estimates. *Med. Biol. Eng. Comput.* **1994**, *32*, 35–42. [CrossRef]
48. Grech, R.; Cassar, T.; Muscat, J.; Camilleri, K.P.; Fabri, S.G.; Zervakis, M.; Xanthopoulos, P.; Sakkalis, V.; Vanrumste, B. Review on solving the inverse problem in EEG source analysis. *J. Neuroeng. Rehabil.* **2008**, *5*, 25. [CrossRef]
49. Pascual-Marqui, R.D. Standardized low-resolution brain electromagnetic tomography (sLORETA): Technical details. *Methods Find. Exp. Clin. Pharmacol.* **2002**, *24*, 5–12.
50. Mulert, C.; Jäger, L.; Schmitt, R.; Bussfeld, P.; Pogarell, O.; Möller, H.J.; Juckel, G.; Hegerl, U. Integration of fMRI and simultaneous EEG: towards a comprehensive understanding of localization and time-course of brain activity in target detection. *NeuroImage* **2004**, *22*, 83–94. [CrossRef]
51. Zumsteg, D.; Friedman, A.; Wieser, H.G.; Wennberg, R.A. Propagation of interictal discharges in temporal lobe epilepsy: correlation of spatiotemporal mapping with intracranial foramen ovale electrode recordings. *Clin. Neurophysiol.* **2006**, *117*, 2615–2626. [CrossRef]
52. Olbrich, S.; Mulert, C.; Karch, S.; Trenner, M.; Leicht, G.; Pogarell, O.; Hegerl, U. EEG-vigilance and BOLD effect during simultaneous EEG/fMRI measurement. *NeuroImage* **2009**, *45*, 319–332. [CrossRef]
53. Pascual-Marqui, R.D.; Lehmann, D.; Koukkou, M.; Kochi, K.; Anderer, P.; Saletu, B.; Tanaka, H.; Hirata, K.; John, E.R.; Prichep, L.; et al. Assessing interactions in the brain with exact low-resolution electromagnetic tomography. *Philos. Trans. R. Soc. Math. Eng. Sci.* **2011**, *369*, 3768–3784. [CrossRef]
54. Lopes, M.A.; Junges, L.; Tait, L.; Terry, J.R.; Abela, E.; Richardson, M.P.; Goodfellow, M. Computational modelling in source space from scalp EEG to inform presurgical evaluation of epilepsy. *Clin. Neurophysiol.* **2020**, *131*, 225–234. [CrossRef] [PubMed]
55. Pascual-Marqui, R.D.; Esslen, M.; Kochi, K.; Lehmann, D. Functional imaging with low-resolution brain electromagnetic tomography (LORETA): A review. *Methods Find. Exp. Clin. Pharmacol.* **2002**, *24*, 91–95. [PubMed]
56. Pisarchik, A.; Jaimes-Reátegui, R.; Villalobos-Salazar, J.; Garcia-Lopez, J.; Boccaletti, S. Synchronization of chaotic systems with coexisting attractors. *Phys. Rev. Lett.* **2006**, *96*, 244102. [CrossRef] [PubMed]
57. Tzourio-Mazoyer, N.; Landeau, B.; Papathanassiou, D.; Crivello, F.; Etard, O.; Delcroix, N.; Mazoyer, B.; Joliot, M. Automated anatomical labeling of activations in SPM using a macroscopic anatomical parcellation of the MNI MRI single-subject brain. *NeuroImage* **2002**, *15*, 273–289. [CrossRef]

Article

Kernel-Based Phase Transfer Entropy with Enhanced Feature Relevance Analysis for Brain Computer Interfaces

Iván De La Pava Panche [1,*], Andrés Álvarez-Meza [2], Paula Marcela Herrera Gómez [3], David Cárdenas-Peña [1], Jorge Iván Ríos Patiño [1] and Álvaro Orozco-Gutiérrez [1]

[1] Automatic Research Group, Universidad Tecnológica de Pereira, Pereira 660003, Colombia; dcardenasp@utp.edu.co (D.C.-P.); jirios@utp.edu.co (J.I.R.P.); aaog@utp.edu.co (Á.O.-G.)
[2] Signal Processing and Recognition Group, Universidad Nacional de Colombia, Manizales 170003, Colombia; amalvarezme@unal.edu.co
[3] Psychiatry, Neuroscience and Community Research Group, Universidad Tecnológica de Pereira, Pereira 660003, Colombia; p.herrera@utp.edu.co
* Correspondence: ide@utp.edu.co

Citation: De La Pava Panche, I.; Álvarez-Meza, A.; Herrera Gómez, P.M.; Cárdenas-Peña, D.; Ríos Patiño, J.I.; Orozco-Gutiérrez, Á. Kernel-Based Phase Transfer Entropy with Enhanced Feature Relevance Analysis for Brain Computer Interfaces. *Appl. Sci.* **2021**, *11*, 6689. https://doi.org/10.3390/app11156689

Academic Editor: Gabriele Cervino

Received: 2 June 2021
Accepted: 19 July 2021
Published: 21 July 2021

Publisher's Note: MDPI stays neutral with regard to jurisdictional claims in published maps and institutional affiliations.

Copyright: © 2021 by the authors. Licensee MDPI, Basel, Switzerland. This article is an open access article distributed under the terms and conditions of the Creative Commons Attribution (CC BY) license (https://creativecommons.org/licenses/by/4.0/).

Abstract: Neural oscillations are present in the brain at different spatial and temporal scales, and they are linked to several cognitive functions. Furthermore, the information carried by their phases is fundamental for the coordination of anatomically distributed processing in the brain. The concept of phase transfer entropy refers to an information theory-based measure of directed connectivity among neural oscillations that allows studying such distributed processes. Phase TE is commonly obtained from probability estimations carried out over data from multiple trials, which bars its use as a characterization strategy in brain–computer interfaces. In this work, we propose a novel methodology to estimate TE between single pairs of instantaneous phase time series. Our approach combines a kernel-based TE estimator defined in terms of Renyi's α entropy, which sidesteps the need for probability distribution computation with phase time series obtained by complex filtering the neural signals. Besides, a kernel-alignment-based relevance analysis is added to highlight relevant features from effective connectivity-based representation supporting further classification stages in EEG-based brain–computer interface systems. Our proposal is tested on simulated coupled data and two publicly available databases containing EEG signals recorded under motor imagery and visual working memory paradigms. Attained results demonstrate how the introduced effective connectivity succeeds in detecting the interactions present in the data for the former, with statistically significant results around the frequencies of interest. It also reflects differences in coupling strength, is robust to realistic noise and signal mixing levels, and captures bidirectional interactions of localized frequency content. Obtained results for the motor imagery and working memory databases show that our approach, combined with the relevance analysis strategy, codes discriminant spatial and frequency-dependent patterns for the different conditions in each experimental paradigm, with classification performances that do well in comparison with those of alternative methods of similar nature.

Keywords: transfer entropy; kernel methods; Renyi's entropy; connectivity analysis; phase interactions

1. Introduction

Neural oscillations are observed in the mammalian brain at different temporal and spatial scales [1]. Oscillations in specific frequency bands are present in distinct neural networks, and their interactions have been linked to fundamental cognitive processes such as attention and memory [2,3] and to information processing at large [4]. Three properties characterize such oscillations: amplitude, frequency, and phase, the latter referring to the position of a signal within an oscillation cycle [5]. Oscillation amplitudes are related to neural synchrony expansion in a local assembly, while the relationships between the phases of neural oscillations, such as phase synchronization, are involved in the coordination of anatomically distributed processing [6]. Moreover, from a functional perspective, phase synchronization and amplitude correlations are independent phenomena [7], hence the interest in studying phase-based interactions independently from other spectral relationships. Additionally, phase relationships are linked to neural synchronization and information flow within networks of connected neural assemblies [8]. Therefore, a measure that aims to capture phase-based interactions among signals from distributed brain regions should ideally include a description of the direction of interaction. A fitting framework for such measure is that of brain effective connectivity [9].

Effective brain connectivity, also known as directed functional connectivity, measures the influence that a neural assembly has over another one, establishing a direction for their interaction by estimating statistical causation from their signals [10]. Directed interactions between oscillations of similar frequency can be captured through measures such as Geweke-Granger causality statistics, partially directed coherence, and directed transfer function [9,11]. However, since these metrics depend on both amplitude and phase signal components, they do not identify phase-specific information flow [8]. The phase slope index (PSI), introduced in [12], measures the direction of coupling between oscillations from the slope of their phases; still, it only captures linear phase relationships [13]. In this context arises the concept of phase transfer entropy, a phase-specific nonlinear directed connectivity measure introduced in [8]. Transfer entropy (TE) is an information-theoretic quantity, based on Wiener's definition of causality, that estimates the directed interaction, or information flow, between two dynamical systems [14,15]. In [8], the authors first extract instantaneous phase time series by complex filtering the signals of interest in a particular frequency, since a signal's phase is only physically meaningful when its spectrum is narrow-banded [16]. Such filtering-based approach has also been explored to obtain phase-specific versions of other information-theoretic metrics such as permutation entropy and time-delayed mutual information [7,16]. Then, the authors compute TE from the obtained phase time series. Nonetheless, since conventional TE estimators are not well suited for periodical variables, in [8] phase TE estimates are obtained through a binning approach performed over multiple trials simultaneously, in a procedure termed trial collapsing.

Phase TE has found multiple applications in neuroscience, such as gaining insight into reduced levels of consciousness by evaluating brain connectivity [17], analyzing resting-state networks [18], and assessing brain connectivity changes in children diagnosed with attention deficit hyperactivity disorder following neurofeedback training [19]. It has even been used to detect fluctuations in financial markets data [20]. Nonetheless, phase TE, estimated as in [8], cannot be employed as a characterization strategy for brain–computer interfaces (BCI) since they require features extracted on an independent trial basis, i.e., each trial must be associated with a set of features. Effective connectivity measures, such as phase TE, can be used to assess the induced physiological variations in the brain occurring during BCI tasks [21]. Discriminative information may be hidden in the dynamical interactions among spatially separated brain regions that characterization methods commonly employed in BCI are not able detect [22]. This information could be relevant to address issues such as the inefficiency problem in some BCI systems [23]. In that context, authors in [6] applied a binning strategy to estimate single-trial phase TE to set up classification systems for visual attention. Nonetheless, binning estimators for single trial-based estimation of information-theoretic measures exhibit systematic bias [8]. Furthermore, spectrally resolved TE estimation methods that can obtain single-trial TE estimates have been recently proposed in the literature [24,25]. Yet, phase TE is conceptually different from them [25], as they are not phase-specific metrics.

Here, we propose a novel methodology to estimate TE between single pairs of instantaneous phase time series. Our approach combines the kernel-based TE estimator we introduced in [10], with phase time series obtained by convolving neural signals with a Morlet Wavelet. The kernel-based TE estimator expresses TE as a linear combination of Renyi's entropy measures of order α [26,27] and then approximates them through functionals defined on positive definite and infinitely divisible kernel matrices [28]. Its most important property is that it sidesteps the need to obtain the probability distributions underlying the data. Instead, the estimator computes TE directly from kernel matrices that, in turn, capture the similarity relations among data. It is robust to varying noise levels and data sizes and to the presence of multiple interaction delays in a network [10]. In this work, we hypothesize that the above-described estimator could overcome the hurdles other single-trial TE estimators face when obtaining TE values from instantaneous phase time series since it would not have to explicitly obtain probability distributions from circular variables [8]. Additionally, since our primary motivation to introduce a robust phase TE estimation methodology is the use of such measures in the context of BCI applications, we also explore a relevance analysis strategy based on centered kernel alignment (CKA) [29]. The CKA-based analysis allows us to identify the set of pairwise channel connectivities relevant to discriminate between specific conditions, favoring the neurophysiological interpretation of our results and providing an option to avoid carrying out all to all channel connectivity estimations in practical BCI systems based on phase TE.

We employ simulated and real-world EEG data to test the introduced effective connectivity measure. The simulated data are obtained from neural mass models, mathematical models of neural mechanisms that generate time series with oscillatory behavior similar to electrophysiological signals. Obtained results for such data show that the proposed kernel-based phase TE estimation method successfully detects the direction of interaction imposed by the model. Indeed, it detects statistically significant connections in the frequency bands of interest, even for weak couplings and narrowband bidirectional interactions. It also displays robustness to realistic levels of noise and signal mixing. Regarding the EEG data, we consider two databases containing signals recorded under two different cognitive paradigms, consisting of motor imagery tasks and a change detection task designed to study working memory. Attained classification results demonstrate that our approach is competitive compared to real-valued and phase-based directed connectivity measures. Thus, this proposal extends the approach described in [10] by introducing a measure that captures directed interactions between the phases of oscillations at specific frequencies. Unlike alternative approaches in the literature, it can be obtained from single trial data, which allows it to be used as a characterization strategy in BCI applications. In addition, the results obtained for the EEG data show that our approach, coupled with the CKA-based relevance analysis, largely outperforms the real-valued kernel-based transfer entropy in [10] as characterization strategy for cognitive tasks such as working memory.

The remainder of the paper is organized as follows: in Section 2 we formally introduce the concept of phase TE and our kernel-based approach for single-trial phase TE estimation. We also describe the proposed CKA-based relevance analysis. Section 3 details the experiments we carried out using simulated and real EEG data in order to evaluate the performance of our proposal. In Section 4 we present and discuss our results, and finally, Section 5 contains our conclusions.

2. Methods
2.1. Phase Transfer Entropy

Transfer entropy (TE) is a Wiener-causal measure of directed interactions between two dynamical systems [14,15]. Given two time series $\mathbf{x} = \{x_t\}_{t=1}^T$ and $\mathbf{y} = \{y_t\}_{t=1}^T$, with $t \in \mathbb{N}$ a discrete time index, $T \in \mathbb{N}$, the TE from $\mathbf{x}$ to $\mathbf{y}$ estimates whether the ability to predict the future of $\mathbf{y}$ improves by considering the past of both $\mathbf{x}$ and $\mathbf{y}$, as compared to the case when only the past of $\mathbf{y}$ is considered. Formally, TE can be defined as:

$$TE(\mathbf{x} \to \mathbf{y}) = \sum_{y_t, \mathbf{y}_{t-1}^{dy}, \mathbf{x}_{t-u}^{dx}} p\left(y_t, \mathbf{y}_{t-1}^{dy}, \mathbf{x}_{t-u}^{dx}\right) \log \left(\frac{p\left(y_t | \mathbf{y}_{t-1}^{dy}, \mathbf{x}_{t-u}^{dx}\right)}{p\left(y_t | \mathbf{y}_{t-1}^{dy}\right)} \right), \quad (1)$$

where $\mathbf{x}_t^{dx}, \mathbf{y}_t^{dy} \in \mathbb{R}^{D \times d}$ are time embedded versions of $\mathbf{x}$ and $\mathbf{y}$, $D = T - (\tau(d-1))$ with $d, \tau \in \mathbb{N}$ the embedding dimension and delay, respectively; $u \in \mathbb{N}$ represents the interaction delay between the driving and the driven systems, and $p(\cdot)$ indicates a probability density function [30] (Henceforth, the summation symbol is to be interpreted in an extended way, that is to say, as a summation or an integral depending on whether the variable is discrete or continuous). Regarding the time embeddings, we have that $\mathbf{x}_t^d = (x(t), x(t-\tau), x(t-2\tau), \ldots, x(t-(d-1)\tau))$ [31,32]. Furthermore, using the definition of Shannon entropy, $H_S(X) = -\sum_x p(x) \log(p(x))$, where X is a discrete random variable ($x \in X$), we can also express Equation (1) as:

$$TE(\mathbf{x} \to \mathbf{y}) = H_S\left(\mathbf{y}_{t-1}^{dy}, \mathbf{x}_{t-u}^{dx}\right) - H_S\left(y_t, \mathbf{y}_{t-1}^{dy}, \mathbf{x}_{t-u}^{dx}\right) + H_S\left(y_t, \mathbf{y}_{t-1}^{dy}\right) - H_S\left(\mathbf{y}_{t-1}^{dy}\right). \quad (2)$$

where $H_S(\cdot, \cdot)$, and $H_S(\cdot)$ stand for joint and marginal entropies.

In phase TE, the time series $\mathbf{x}$ and $\mathbf{y}$ are replaced by instantaneous phase time series $\theta^x(f) \in [-\pi, \pi]_{t=1}^T$ and $\theta^y(f) \in [-\pi, \pi]_{t=1}^T$, obtained from $\mathbf{s}_x = \varsigma_x e^{i\theta^x(f)} \in \mathbb{C}^T$ and $\mathbf{s}_y = \varsigma_y e^{i\theta^y(f)} \in \mathbb{C}^T$, which contain the complex-filtered values of $\mathbf{x}$ and $\mathbf{y}$ at frequency f, respectively, and with $\varsigma_x, \varsigma_y \in \mathbb{R}^T$ the amplitude envelopes of the filtered time series [8]. Thus, we have that

$$TE^\theta(\mathbf{x} \to \mathbf{y}, f) = H_S\left(\theta_{t-1}^{y,dy}, \theta_{t-u}^{x,dx}\right) - H_S\left(\theta_t^y, \theta_{t-1}^{y,dy}, \theta_{t-u}^{x,dx}\right) + H_S\left(\theta_t^y, \theta_{t-1}^{y,dy}\right) - H_S\left(\theta_{t-1}^{y,dy}\right), \quad (3)$$

where $\theta_t^{x,dx}$ and $\theta_t^{y,dy}$ are time embedded versions of θ^x and θ^y. Note that for the sake of notation simplicity we have dropped the explicit dependency of the phase time series on f.

2.2. Kernel-Based Renyi's Phase Transfer Entropy

In [10] we propose a TE estimator based on kernel matrices that approximate Renyi's entropy measures of order α. This data-driven approach has the advantage of sidestepping the need for probability distribution estimation in TE computation. First, we show that TE can be expressed as

$$TE_\alpha(\mathbf{x} \to \mathbf{y}) = H_\alpha\left(\mathbf{y}_{t-1}^{dy}, \mathbf{x}_{t-u}^{dx}\right) - H_\alpha\left(y_t, \mathbf{y}_{t-1}^{dy}, \mathbf{x}_{t-u}^{dx}\right) + H_\alpha\left(y_t, \mathbf{y}_{t-1}^{dy}\right) - H_\alpha\left(\mathbf{y}_{t-1}^{dy}\right). \quad (4)$$

where $H_\alpha(X)$ stands for Renyi's α entropy, a generalization of Shannon's entropy [26,27], defined as

$$H_\alpha(X) = \frac{1}{1-\alpha} \log\left(\sum_x p(x)^\alpha dx\right), \quad (5)$$

with $\alpha \neq 1$ and $\alpha \geq 0$. In the limiting case where $\alpha \to 1$, it tends to Shannon's entropy. Then, using the kernel-based formulation for Renyi's α entropy introduced in [28],

$$H_\alpha(\mathbf{A}) = \frac{1}{1-\alpha} \log(\text{tr}(\mathbf{A}^\alpha)), \quad (6)$$

where $\mathbf{A} \in \mathbb{R}^{n \times n}$ is a Gram matrix with elements $a_{ij} = \kappa(x_i, x_j)$, $\kappa(\cdot, \cdot) \in \mathbb{R}$ stands for a positive definite and infinitely divisible kernel function, n for the number of realizations of X, and $\text{tr}(\cdot)$ for the matrix trace; along with the accompanying formulation for the Renyi's α entropy of joint probability distributions,

$$H_\alpha(\mathbf{A}, \mathbf{B}) = H_\alpha\left(\frac{\mathbf{A} \circ \mathbf{B}}{\text{tr}(\mathbf{A} \circ \mathbf{B})}\right) = \frac{1}{1-\alpha} \log\left(\text{tr}\left(\left(\frac{\mathbf{A} \circ \mathbf{B}}{\text{tr}(\mathbf{A} \circ \mathbf{B})}\right)^\alpha\right)\right), \quad (7)$$

where $\mathbf{B} \in \mathbb{R}^{n \times n}$ is a second Gram matrix and the operator $\circ$ stands for the Hadamard product, we estimate the TE_α from $\mathbf{x}$ to $\mathbf{y}$ as:

$$TE_{\kappa\alpha}(\mathbf{x} \to \mathbf{y}) = H_\alpha\left(\mathbf{K}_{\mathbf{y}_{t-1}^{dy}}, \mathbf{K}_{\mathbf{x}_{t-u}^{dx}}\right) - H_\alpha\left(\mathbf{K}_{y_t}, \mathbf{K}_{\mathbf{y}_{t-1}^{dy}}, \mathbf{K}_{\mathbf{x}_{t-u}^{dx}}\right)$$
$$+ H_\alpha\left(\mathbf{K}_{y_t}, \mathbf{K}_{\mathbf{y}_{t-1}^{dy}}\right) - H_\alpha\left(\mathbf{K}_{\mathbf{y}_{t-1}^{dy}}\right), \quad (8)$$

where the kernel matrices $\mathbf{K}_{\mathbf{y}_t}, \mathbf{K}_{\mathbf{y}_{t-1}^{dy}}, \mathbf{K}_{\mathbf{x}_{t-u}^{dx}} \in \mathbb{R}^{(D-u) \times (D-u)}$ hold elements $k_{ij} = \kappa(\mathbf{a}_i, \mathbf{a}_j)$. For $\mathbf{K}_{\mathbf{y}_t}$, $a_i, a_j \in \mathbb{R}$ are the values of the time series $\mathbf{y}$ at times i and j. While for $\mathbf{K}_{\mathbf{y}_{t-1}^{dy}}$, the vectors $\mathbf{a}_i, \mathbf{a}_j \in \mathbb{R}^d$ contain the time embedded version of $\mathbf{y}$, $\mathbf{y}_t^{dy}$, at times i and j, adjusted according to the time indexing of TE. The same logic holds true for $\mathbf{K}_{\mathbf{x}_{t-u}^{dx}}$.

In this study, we hypothesize that the above-described TE estimator, having previously displayed robustness to common issues that affect connectivity analyses [10], could overcome many of the problems associated with single-trial phase TE estimation [8]. Hence, we propose a kernel-based Renyi's phase TE estimator defined as:

$$TE_{\kappa\alpha}^{\theta}(\mathbf{x} \to \mathbf{y}, f) = H_\alpha\left(\mathbf{K}_{\theta_{t-1}^{y,dy}}, \mathbf{K}_{\theta_{t-u}^{x,dx}}\right) - H_\alpha\left(\mathbf{K}_{\theta_t}, \mathbf{K}_{\theta_{t-1}^{y,dy}}, \mathbf{K}_{\theta_{t-u}^{x,dx}}\right) + H_\alpha\left(\mathbf{K}_{\theta_t}, \mathbf{K}_{\theta_{t-1}^{y,dy}}\right) - H_\alpha\left(\mathbf{K}_{\theta_{t-1}^{y,dy}}\right), \quad (9)$$

where the kernel matrices $\mathbf{K}_{\theta_t}, \mathbf{K}_{\theta_{t-1}^{y,dy}}, \mathbf{K}_{\theta_{t-u}^{x,dx}} \in \mathbb{R}^{(D-u) \times (D-u)}$ hold elements analogous to those of matrices $\mathbf{K}_{\mathbf{y}_t}$, $\mathbf{K}_{\mathbf{y}_{t-1}^{dy}}$, and $\mathbf{K}_{\mathbf{x}_{t-u}^{dx}}$ in Equation (8), while replacing the time series $\mathbf{x}$ and $\mathbf{y}$ for their instantaneous phase time series θ^x and θ^y at frequency f, respectively.

2.3. Phase-Based Effective Connectivity Estimation Approaches Considered in This Study

2.3.1. Phase Transfer Entropy

We obtain phase TE values through three different estimators that allow computing TE from individual signal pairs. First, the proposed kernel-based Renyi's phase TE estimator ($TE_{\kappa\alpha}^{\theta}$), defined in Equation (9). Second, the Kraskov-Stögbauer-Grassberger TE estimator (TE_{KSG}^{θ}), a method that relies on a local approximation of the probability distributions needed to estimate the entropies in TE from the distances of every data point to its neighbors [33,34]. Thirdly, an approach termed symbolic TE (TE_{Sym}^{θ}) that relies on a symbolization scheme based on ordinal patterns. The symbolization scheme allows estimating the probabilities involved in the computation of TE directly from the symbols' relative frequencies [35].

In all cases, θ^x and θ^y are obtained by convolving the real-valued time series with a Morlet wavelet, defined as

$$h(t, f) = \exp(-t^2/2\xi_t^2)\exp(i2\pi f t), \quad (10)$$

where f stands for the filter frequency, $\xi_t = m/2\pi f$ is the time domain standard deviation of the wavelet, and m defines the compromise between time and frequency resolution [8].

2.3.2. Phase Slope Index

The phase slope index (PSI) is an effective brain connectivity measure that assesses the direction of coupling between two oscillatory signals of similar frequencies [13]. Given two time series $\mathbf{x} = \{x_i\}_{t=1}^l$ and $\mathbf{y} = \{y_i\}_{t=1}^l$, the PSI is defined as the slope of the phase of the cross-spectra between $\mathbf{x}$ and $\mathbf{y}$:

$$PSI(\mathbf{x} \to \mathbf{y}) = \Im\left(\sum_{f \in F} C_{\mathbf{xy}}^*(f) C_{\mathbf{xy}}(f + df)\right), \quad (11)$$

where $C_{\mathbf{xy}}(f) = S_{\mathbf{xy}}/\sqrt{S_{\mathbf{xx}} S_{\mathbf{yy}}}$ is the complex coherence, $S_{\mathbf{xy}} \in \mathbb{C}$ is the cross-spectrum between $\mathbf{x}$ and $\mathbf{y}$, $S_{\mathbf{xx}}, S_{\mathbf{yy}} \in \mathbb{C}$ are the auto-spectrums of $\mathbf{x}$ and $\mathbf{y}$, $df \in \mathbb{R}^+$ is the frequency resolution, F stands for the set of frequencies over which the slope is summed, and $\Im$ indicates selecting only the imaginary part of the sum [12]. If the PSI, as defined in Equation (11), is positive, then there is directed interaction from $\mathbf{x}$ to $\mathbf{y}$ in F. Conversely, if the PSI is negative, the directed interaction goes from $\mathbf{y}$ to $\mathbf{x}$. Note that by definition the PSI is an antisymmetric measure: $PSI(\mathbf{x} \to \mathbf{y}) = -PSI(\mathbf{y} \to \mathbf{x})$.

2.3.3. Granger Causality

We also characterize the simulated and EEG data using Granger causality (GC). Like TE, GC is derived from Wiener's definition of causality, and the two measures, in their original forms, are equivalent for Gaussian variables [36]. Briefly, for two stationary time series $\mathbf{x} = \{x_t\}_{t=1}^{T}$ and $\mathbf{y} = \{y_i\}_{t=1}^{T}$, the Granger causality from $\mathbf{x}$ to $\mathbf{y}$ is defined as:

$$\mathrm{GC}(\mathbf{x} \to \mathbf{y}) = \log\left(\frac{\mathrm{var}\{\mathbf{e}\}}{\mathrm{var}\{\mathbf{e}'\}}\right), \tag{12}$$

where $\mathbf{e}, \mathbf{e}' \in \mathbb{R}^{T-o}$ are vectors holding the residual or prediction errors of two autoregressive models, and $\mathrm{var}\{\cdot\}$ stands for the variance operator. The errors in $\mathbf{e}$ come from an autoregressive process of order o that predicts $\mathbf{y}$ from its past values alone. On the other hand, the errors in $\mathbf{e}'$ come from a bivariate autoregressive process of order o that predicts $\mathbf{y}$ from the past values of $\mathbf{y}$ and $\mathbf{x}$ [11]. If the past of $\mathbf{x}$ improves the prediction of $\mathbf{y}$ then $\mathrm{var}\{\mathbf{e}\} \gg \mathrm{var}\{\mathbf{e}'\}$ and $\mathrm{GC}(\mathbf{x} \to \mathbf{y}) \gg 0$, if it does not, then $\mathrm{var}\{\mathbf{e}\} \approx \mathrm{var}\{\mathbf{e}'\}$ and $\mathrm{GC}(\mathbf{x} \to \mathbf{y}) \to 0$. In addition, in analogy to the concept of phase TE, we define $\mathrm{GC}^\theta(\mathbf{x} \to \mathbf{y}, f) = \mathrm{GC}(\theta^x \to \theta^y)$, where θ^x and θ^y are instantaneous phase time series obtained by filtering $\mathbf{x}$ and $\mathbf{y}$ at frequency f, as a measure within the framework of GC that captures phase-based interactions.

2.4. Kernel-Based Relevance Analysis

When characterizing EEG data through effective brain connectivity measures for BCI-related applications, two common and related issues can arise. First, all to all channel connectivity analyses result in a large number of features, many of which may not provide useful information to discriminate between the conditions of the BCI paradigm of interest [10]. This can add noise and complexity to any subsequent analysis stage. Second, estimating such a large number of pairwise channel connectivities can be computationally expensive, especially for measures such as TE and single-trial TE$^\theta$ [8], which can hinder their inclusion in practical BCI systems. Both problems could be addressed by identifying the set of pairwise channel connectivities that are relevant to discriminate between specific conditions, which would also lead to a clearer neurophysiological interpretation of the obtained results [6,10]. To that end, we explore a relevance analysis strategy based on centered kernel alignment (CKA). CKA allows quantifying the similarity between two sample spaces by comparing two characterizing kernel functions [29]. First, assume we have a feature matrix $\boldsymbol{\Phi} \in \mathbb{R}^{N \times P}$, and a corresponding vector of labels $\mathbf{l} \in \mathbb{Z}^N$, with N the number of observations and P the number of features. For the case of connectivity-based EEG analysis, each element in $\boldsymbol{\Phi}$ holds a connectivity value for a pair of channels, with each row of $\boldsymbol{\Phi}$ containing multiple connectivity values (features) estimated for a single trial or observation. The corresponding element in $\mathbf{l}$ holds a label identifying the condition associated to that trial. Next, we define two kernel matrices $\mathbf{K}_{\boldsymbol{\Phi}} \in \mathbb{R}^{N \times N}$ and $\mathbf{K}_l \in \mathbb{R}^{N \times N}$. The first matrix holds elements $k_{ij}^{\boldsymbol{\Phi}} = \kappa_{\boldsymbol{\Phi}}(\boldsymbol{\phi}_i, \boldsymbol{\phi}_j)$ with $\boldsymbol{\phi}_i, \boldsymbol{\phi}_j \in \mathbb{R}^P$ row vectors belonging to $\boldsymbol{\Phi}$, and

$$\kappa_{\boldsymbol{\Phi}}(\boldsymbol{\phi}_i, \boldsymbol{\phi}_j; \sigma) = \exp\left(-\frac{d^2(\boldsymbol{\phi}_i, \boldsymbol{\phi}_j)}{2\sigma^2}\right) \tag{13}$$

a radial basis function (RBF) kernel [37], where $d^2(\cdot, \cdot)$ is a distance operator and $\sigma \in \mathbb{R}^+$ is the kernel's bandwidth. The second matrix has elements $k_{ij}^l = \kappa_l(l_i, l_j)$ with $l_i, l_j \in \mathbf{l}$, and

$$\kappa_l(l_i, l_j) = \delta(l_i - l_j), \tag{14}$$

a dirac kernel, where $\delta(\cdot)$ stands for the Dirac delta. Then, the CKA can be estimated as:

$$\hat{\rho}(\bar{\mathbf{K}}_{\boldsymbol{\Phi}}, \bar{\mathbf{K}}_l) = \frac{\langle \bar{\mathbf{K}}_{\boldsymbol{\Phi}}, \bar{\mathbf{K}}_l \rangle_F}{(\langle \bar{\mathbf{K}}_{\boldsymbol{\Phi}}, \bar{\mathbf{K}}_{\boldsymbol{\Phi}} \rangle_F \langle \bar{\mathbf{K}}_l, \bar{\mathbf{K}}_l \rangle_F)^{1/2}}, \tag{15}$$

where $\bar{K} \in \mathbb{R}^{N \times N}$ is the centered version of K, obtained as $\bar{K} = \tilde{I} K \tilde{I}$, where $\tilde{I} = I - \mathbf{1}^\top \mathbf{1}/N$ is the empirical centering matrix, $I \in \mathbb{R}^{N \times N}$ is the identity matrix, $\mathbf{1} \in \mathbb{R}^N$ is an all-ones vector, and $\langle \bar{K}, \bar{K} \rangle_F = \sqrt{\text{tr}(\bar{K}\bar{K}^T)}$ denotes the matrix-based Frobenius norm. Now, for κ_Φ we select as distance operator the the Mahalanobis distance

$$d_A^2(\phi_i, \phi_j) = (\phi_i - \phi_j) \Gamma \Gamma^\top (\phi_i - \phi_j)^\top \tag{16}$$

where $\Gamma \in \mathbb{R}^{P \times Q}$, $Q \leq P$, is a linear projection matrix, and $\Gamma \Gamma^\top$ is the corresponding inverse covariance matrix. Afterward, the projection matrix Γ is obtained by solving the following optimization problem:

$$\hat{\Gamma} = \arg\max_\Gamma \log(\hat{\rho}(\bar{K}_\Phi, \bar{K}_l; \Gamma)), \tag{17}$$

where the logarithm function is used for mathematical convenience. $\hat{\Gamma}$ can be estimated through standard stochastic gradient descent, as detailed in [38], through the update rule

$$\Gamma^{r+1} = \Gamma^r - \mu_\Gamma^r \nabla_{\Gamma^r}(\hat{\rho}(K_\Phi, K_l)), \tag{18}$$

where $\mu \in \mathbb{R}^+$ is the step size of the learning rule, and r indicates a time step. Finally, we quantify the contribution of each feature to the projection matrix $\hat{\Gamma}$, which maximizes the alignment between the feature and label spaces, by building a relevance vector index $\varrho \in \mathbb{R}^P$, whose elements are defined as:

$$\varrho_p = \sum_{q=1}^{Q} |\gamma_{pq}|; \forall p \in P, \quad \gamma \in \Gamma. \tag{19}$$

ϱ can then be used to rank the features in Φ according to their discrimination capability. A high ϱ_p value indicates that the p-th feature in Φ, in our case a connection between a specific pair of channels, is relevant when it comes to distinguishing between the conditions contained in the label vector l.

3. Experiments

In order to test the performance of our single-trial phase TE estimator we carry out experiments on simulated data from neural mass models, and on real EEG data, obtained under motor imagery and visual working memory paradigms. We then compare our results with those obtained with the alternative approaches for phase-based effective connectivity estimation detailed in Section 2.3.

3.1. Neural Mass Models

Neural mass models (NMM) are biologically plausible mathematical descriptions of neural mechanisms [39]. They represent the electrical activity of neural populations at a macroscopic level through a set of stochastic differential equations [40]. NMMs allow generating mildly nonlinear time series with properties that resemble the oscillatory dynamics of electrophysiological signals, such as EEG, and how they change as a result of coupling between different cortical areas. Therefore, NMMs are useful to study the behavior of brain connectivity measures that aim to capture such interactions [8,24,40,41]. Figure 1A shows a schematic representation of an NMM with two interacting cortical areas from which two signals, **x** and **y** (see Figure 1B), can be obtained. The parameters C_{12} and C_{21} are known as coupling coefficients, and they determine the strength of the coupling from Area 1 to Area 2, and from Area 2 to Area 1, respectively. The parameter ν represents the interaction lag between the two areas, while $\mathbf{p}_1$ and $\mathbf{p}_2$ are external inputs coming from other cortical regions.

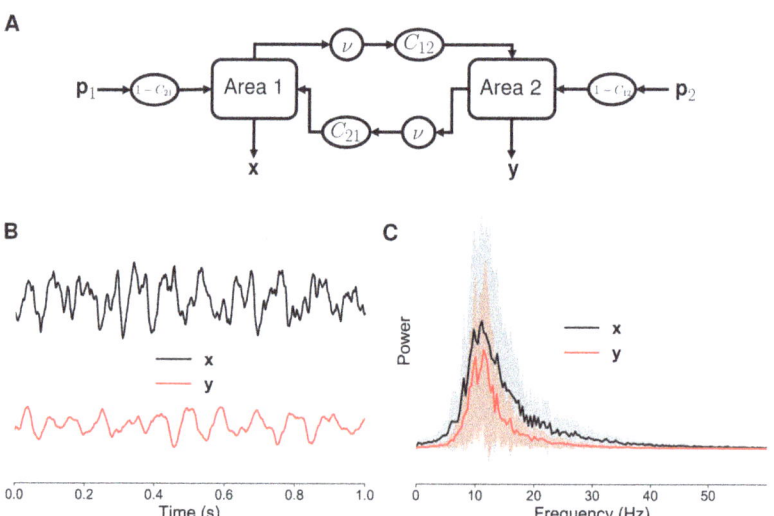

Figure 1. (**A**) Schematic representation of a neural mass model. (**B**) 1 s long unidirectionally coupled time series generated by the model. (**C**) Average power spectrums peaking in the α and lower β frequency bands.

In this work, we use NMMs to generate interacting time series with known oscillatory properties in order to test the performance of the proposed phase TE estimator. In particular, we test our proposal in terms of its ability to detect directed interactions for different levels of coupling strength, under the presence of noise and signal mixing, and for bidirectional narrowband couplings. We proceed as follows: first, we set the model parameters describing Areas 1 and 2 as in [40], so as to generate signals with power spectrums peaking in the α (8 Hz–12 Hz) and lower β bands (12–20 Hz), as depicted in Figure 1C. Then, in order to generate unidirectionally coupled signals, with interactions from **x** to **y**, we set the parameter C_{21} to 0 for all simulations. Also, the parameter ν is set to 20 ms, and the extrinsic inputs $\mathbf{p}_1$ and $\mathbf{p}_2$ are modeled as Gaussian noise [40]. Afterward, we generate 50 pairs (trials) of 3 s long signals, using a simulation time step of 1 ms, equivalent to a sampling frequency of 1000 Hz, for each condition in the three scenarios detailed in Sections 3.1.1–3.1.3. Next, we select a 2 s long segment from the signals, from 0.5 s to 2.5 s, and downsample them to 250 Hz. Then, we compute connectivity estimates for the simulated data in the frequency range between 2 Hz and 60 Hz, in steps of 2 Hz. After that, we obtain net connectivity values, defined as

$$\Delta\lambda(\mathbf{x},\mathbf{y},f) = \lambda(\mathbf{x} \to \mathbf{y},f) - \lambda(\mathbf{y} \to \mathbf{x},f), \qquad (20)$$

where λ stands for any of the phase-based effective connectivity measures studied, except for the PSI, in which case all subsequent analyses are performed directly on the PSI values. Lastly, for each condition in the three scenarios and at each frequency evaluated, we perform permutation tests based on randomized surrogate trials [34,42] to determine which net couplings or directed connections are statistically significant. The permutation test employed uses the trial structure of the data to generate surrogate datasets for the null hypothesis (absence of directed interactions). It does so by shuffling the data from different trials. The significance level for the tests was set to 3.3×10^{-4} after applying the Bonferroni correction to an initial alpha level of 0.01 in order to account for 30 independent tests, one for each evaluated frequency per condition.

3.1.1. Coupling Strength

In order to test the ability of our phase TE estimation method to detect phase-based directed interactions of varying intensity, we modify the coupling strength between the simulated signals, **x** and **y**, by varying the parameter C_{12} in the range $\{0, 0.2, 0.5, 0.8\}$, with 0 indicating the absence of coupling and 0.8 a strong interaction between the two signals.

3.1.2. Noise and Signal Mixing

To asses the robustness of our proposal to realistic levels of noise and signal mixing, we do the following: we generate a noise time series η, with the same power spectrum of **x**, through the methodology proposed in [8]. Then, we add **x** and η to generate a noisy version of **x**, $\mathbf{x}_\eta = \mathbf{x} + 10^{-\frac{SNR}{20}} \eta$, where SNR is the signal to noise ratio. Likewise for **y**. Then, we mix $\mathbf{x}_\eta$ and $\mathbf{y}_\eta$ to simulate one of the effects of volume conduction, by doing $\mathbf{x}_\eta^w = (1 - \frac{w}{2})\mathbf{x}_\eta + (\frac{w}{2})\mathbf{y}_\eta$, and $\mathbf{y}_\eta^w = (1 - \frac{w}{2})\mathbf{y}_\eta + (\frac{w}{2})\mathbf{x}_\eta$, with w the mixing strength. We set the parameters SNR and w to 3 and 0.25 respectively, based on the results obtained in [8] for realistic values of noise and mixing for EEG signals. The coupling coefficient C_{12} is held constant at a value of 0.5 to simulate couplings of medium strength.

3.1.3. Narrowband Bidirectional Interactions

In this experiment, we aim to evaluate how our proposal deals with bidirectional interactions of localized frequency content. Particularly, we want to assess its performance for signals **x** and **y** containing a directed interaction from **x** to **y** at 10 Hz and an interaction in the opposite direction, from **y** to **x**, at 40 Hz. To generate such signals, first, we modify the model parameters of Area 2 so that it produces a signal **y** with a power spectrum peaking in the γ band [39]. The power spectrum of **x** remains as before. The coupling coefficient C_{12} is again held constant at a value of 0.5. The change in the parameters of Area 2 leads to strong directed interactions from **x** to **y** around 10 Hz and 40 Hz. Then, we use a Morlet wavelet (Equation (10)) to filter both **x** and **y** at those frequencies (10 Hz and 40 Hz). The obtained real-valued narrowband time series are then combined as follows: $\mathbf{x}_* = \mathbf{x}_{10\,Hz} + \mathbf{y}_{40\,Hz}$ and $\mathbf{y}_* = \mathbf{y}_{10\,Hz} + \mathbf{x}_{40\,Hz}$. Next, $\mathbf{x}_*$ and $\mathbf{y}_*$ are added to broadband noise generated following the same approach described in Section 3.1.2, with an SNR of 6.

3.2. EEG Data

In order to test the performance of our phase TE estimator in the context of BCI, we obtain effective connectivity features from EEG signals recorded under two different cognitive paradigms: the first one consisting of motor imagery (MI) tasks and the second one of a change detection task designed to study working memory (WM). Our aims are to set up classification systems that allow discriminating between the conditions in each paradigm, using as inputs relevant directed interactions among EEG signals and then evaluate their performance in relation to the connectivity measures used to train them. To those ends, we employ two publicly available databases: the BCI Competition IV database 2a (http://www.bbci.de/competition/iv/index.html, accessed on 2 June 2021) and a database from brain activity during visual working memory (https://data.mendeley.com/datasets/j2v7btchdy/2, accessed on 2 June 2021).

3.2.1. Motor Imagery

Motor imagery (MI) is the process of mentally rehearsing a motor action, such as moving a limb, without actually executing it [43]. The BCI Competition IV database 2a [44] comprises EEG data from 9 healthy subjects recorded during an MI paradigm consisting of four different MI tasks, namely, imagining the movement of the left hand, the right hand, both feet, or the tongue. Each trial of the paradigm starts with a fixation cross displayed on a computer screen, along with a beep. At second 2, a visual cue appears on the screen for a period of 1.25 s (an arrow pointing left, right, down, or up, corresponding to one of the four MI tasks). The cue prompts the subject to perform the indicated MI task until the cross vanishes from the screen at second 6. A representation of the paradigm's time course is shown in Figure 2A. Each subject performed 144 trials per MI task. The EEG data are acquired at a sampling rate of 250 Hz, from 22 Ag/AgCl electrodes ($C = 22$) placed according to the international 10/20 system, as depicted in Figure 2B. Next, the data are bandpass-filtered between 0.5 Hz and 100 Hz. A 50 Hz Notch filter is also applied. For each subject, the database contains a training dataset and a testing dataset, obtained following the same experimental paradigm [44]. In this study, we consider a bi-class classification problem involving the left and right hand MI tasks, so we drop the trials associated with the feet and the tongue. Afterward, we also drop the trials marked for rejection in the database itself [44]. Then, for all trials we select a 2 s long time window stretching from second 3 to second 5 ($M = 500$ samples), as schematized in Figure 2A. Finally, we compute the surface Laplacian of each remaining trial through the spherical spline method for source current density estimation, in order to reduce the deleterious effects of volume conduction on connectivity analyses [21,45,46].

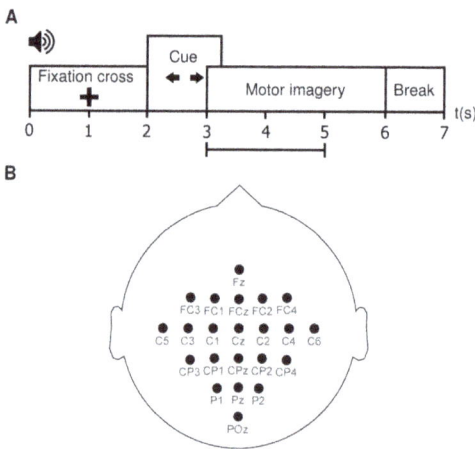

Figure 2. (**A**) Schematic representation of the MI protocol. (**B**) EEG channel montage used for the acquisition of the MI database.

3.2.2. Working Memory

The concept of working memory (WM) refers to a cognitive system of limited capacity that allows for temporary storage and manipulation of information [47]. The database from brain activity during visual working memory, presented in [48], contains EEG data recorded from twenty-three subjects, with normal or corrected-to-normal vision, and without color-vision deficiency, while performing multiple trials of a change detection task [49]. The task consists of remembering the colors of a set of squares, termed memory array, and then comparing them with the colors of a second set of squares located in the same positions, termed test array. A trial of the task begins with an arrow indicating either the left or the right side of the screen. Then, a memory array appears on the screen for 0.1 s. For every trial, memory arrays are displayed on both hemifields, but the subject must remember only those appearing on the side indicated by the arrow cue. Next, after a retention period lasting 0.9 s, a test array appears. The subject then reports if the colors of all the items in the memory and test arrays match. The task has three levels according to the number of elements in the memory array: low memory load (one square), medium memory load (two squares), and high memory load (four squares). A representation of the above-described experimental paradigm is depicted in Figure 3A. The color of one of the squares in the test array differs from its counterpart in the memory array in 50% of the trials. Each subject performed a total of 96 trials, with 32 trials for each memory load level. The EEG data are acquired at a sampling rate of 2048 Hz, using 64 electrodes (Biosemi ActiveTwo) arranged according to the international 10/20 extended system, as depicted in Figure 2B. Besides the

EEG data, the database provides recordings from four EOG channels and two externals electrodes located on the left and right mastoids.

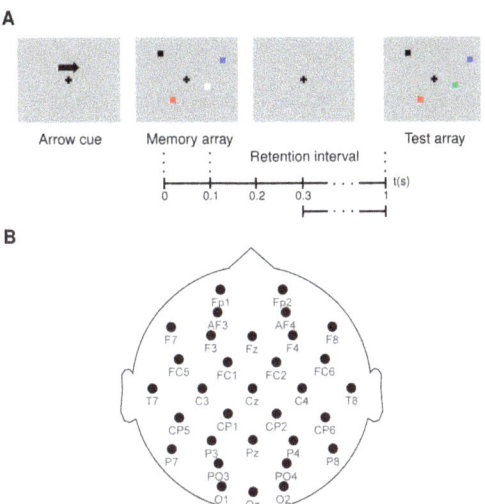

Figure 3. (**A**) Schematic representation of the WM protocol. (**B**) EEG channel montage used for the acquisition of the WM database.

In this study, we perform the following preprocessing steps before any further data analysis. First, we re-reference the data to the average of the mastoid channels. Next, we bandpass-filter the data between 0.01 Hz and 20 Hz using a Butterworth filter of order 2. Afterward, we extract the trial information from the continuous EEG data using a 1.4 s squared window. Each trial segment starts 0.2 s before the presentation of the memory array. Then, we perform a visual inspection of the data and discard two subjects (subjects number 11 and 17) because of the presence of strong artifacts in a very large number of trials. Subjects number 22 and 23 are reassigned as subjects 17 and 11, respectively. After that, we remove ocular artifacts from the EEG data by performing independent component analysis (ICA) on it and then eliminating the components that more closely resemble the information provided by the EOG data [48]. Then, we discard all incorrect trials, i.e., trials for which the subjects incorrectly matched the memory and test arrays. Next, we select 32 out of the 64 channels in the EEG data ($C = 32$), as shown in Figure 3B. Then, we downsample the data to 1024 Hz, and segment, for each trial, the time window starting 0.3 s after the onset of the memory array and ending just before the presentation of the test array (see Figure 3A). The 0.7 s long segments ($M = 717$) cover most of the retention interval, the period when the subjects should maintain the stimulus information in their working memories, while leaving out any purely sensory responses elicited immediately after the presentation of the stimulus. Finally, with the aim of reducing the presence of spurious connections associated with volume conduction effects, we compute the surface Laplacian of each trial.

3.2.3. Classification Setup

Feature Extraction

Let $\Psi = \{\mathbf{X}_n \in \mathbb{R}^{C \times M}\}_{n=1}^{N}$ be an EEG set holding N trials from either an MI or a WM dataset, recorded from a single subject, where C stands for the number of channels and M corresponds to the number of samples. In addition, let $\{l_n\}_{n=1}^{N}$ be a set whose n-th element is the label associated with trial $\mathbf{X}_n$. For the MI database l_n can take the values of 1 and 2, corresponding to right hand and left hand motor imagination, respectively. Similarly, for the WM database, l_n can take the values of 1, 2, and 3 corresponding to low, medium, and high memory loads. In both cases, our goal is to estimate the class label from relevant effective connectivity features extracted from $\mathbf{X}_n$. Because of the results obtained for the simulated data (see Section 4.1 for details), here we consider features from only three approaches for phase-based effective connectivity estimation, namely, $TE_{\kappa\alpha}^{\theta}$, GC^{θ}, and PSI. Additionally, we also characterize the data through the real-valued versions of $TE_{\kappa\alpha}$ and GC.

For the real-valued effective connectivity measures considered, we do the following: let $\lambda(\mathbf{x}_c \rightarrow \mathbf{x}_{c'})$ be a measure of effective connectivity between channels $\mathbf{x}_c, \mathbf{x}_{c'} \in \mathbb{R}^M$. By computing $\lambda(\mathbf{x}_c \rightarrow \mathbf{x}_{c'})$ for each pairwise combination of channels in $\mathbf{X}_n$ we obtain a connectivity matrix $\mathbf{\Lambda} \in \mathbb{R}^{C \times C}$. In the case when $c = c'$, we set $\lambda(\mathbf{x}_c \rightarrow \mathbf{x}_{c'}) = 0$. Then, we normalize $\mathbf{\Lambda}$ to the range $[0, 1]$. After performing the above procedure for the N trials, we get set of connectivity matrices $\{\mathbf{\Lambda}_n \in \mathbb{R}^{C \times C}\}_{n=1}^{N}$. Then, we apply vector concatenation to $\mathbf{\Lambda}_n$ to yield a vector $\boldsymbol{\phi}_n \in \mathbb{R}^{1 \times (C \times C)}$. Next, we stack the N vectors $\boldsymbol{\phi}_n$, corresponding to each trial, to obtain a matrix $\boldsymbol{\Phi} \in \mathbb{R}^{N \times (C \times C)}$ holding all directed interactions, estimated through λ, for the EEG set Ψ. A graphical representation of the above-described steps, as well as of our overall classification setup, is depicted in Figure 4.

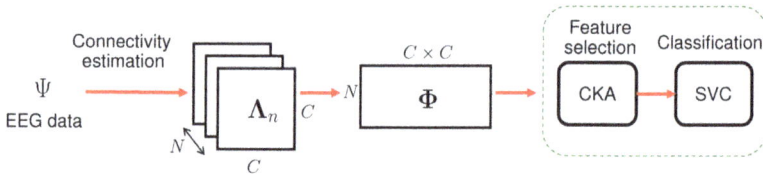

Figure 4. Schematic representation of our overall classification setup.

For the phase-based effective connectivity measures of interest, we proceed in a similar fashion: let $\lambda^{\theta}(\mathbf{x}_c \rightarrow \mathbf{x}_{c'}, f)$ be a measure of phase-based effective connectivity between channels $\mathbf{x}_c, \mathbf{x}_{c'}$ at frequency f. By computing $\lambda^{\theta}(\mathbf{x}_c \rightarrow \mathbf{x}_{c'}, f)$ for each pairwise combination of channels in $\mathbf{X}_n$ we obtain a connectivity matrix $\mathbf{\Lambda}(f) \in \mathbb{R}^{C \times C}$ (when $c = c'$, we set $\lambda^{\theta}(\mathbf{x}_c \rightarrow \mathbf{x}_{c'}, f) = 0$). For the MI database, we vary the values of f in the range from 8 Hz to 18 Hz, in 2 Hz steps, since activity in that frequency range has been associated with MI tasks [43]. Then we define two bandwidths of interest $\Delta f \in \{\alpha \in [8 - 12], \beta_l \in [14 - 18]\}$ Hz. Afterward, we average the matrices $\mathbf{\Lambda}(f)$ within each bandwidth, normalize the resulting matrices to the range $[0, 1]$, and stack them together, so that for each trial we have a connectivity matrix $\mathbf{\Lambda}' \in \mathbb{R}^{C \times C \times 2}$. Therefore, for the N trials, we get set of connectivity matrices $\{\mathbf{\Lambda}'_n \in \mathbb{R}^{C \times C \times 2}\}_{n=1}^{N}$. Then, we apply vector concatenation to $\mathbf{\Lambda}'_n$ to yield a vector $\boldsymbol{\phi}_n \in \mathbb{R}^{1 \times (C \times C \times 2)}$. After that, we stack the N vectors $\boldsymbol{\phi}_n$ in order to obtain a single matrix $\boldsymbol{\Phi} \in \mathbb{R}^{N \times (C \times C \times 2)}$ characterizing Ψ for the MI data. For the WM we follow the same steps, only that in this case we vary the values of f in the range from 4 Hz to 18 Hz, in 2 Hz steps, since oscillatory activity at those frequencies has been shown to play a role in the interactions between different brain regions during WM [50,51]. Next, we define three bandwidths of interest $\Delta f \in \{\theta \in [4 - 6], \alpha \in [8 - 12], \beta_l \in [14 - 18]\}$ Hz, which leads to a connectivity matrix $\mathbf{\Lambda}' \in \mathbb{R}^{C \times C \times 3}$ for each trial and ultimately to a matrix $\boldsymbol{\Phi} \in \mathbb{R}^{N \times (C \times C \times 3)}$ characterizing Ψ for the WM data. Note that since the PSI is an antisymmetric connectivity measure, we only use the upper triangular part of the connectivity matrix associated with each trial to build $\boldsymbol{\Phi}$.

Feature Selection and Classification

After characterizing the EEG data, either through real-valued or phase-based effective connectivity measures, we set up a subject-dependent classification system for the MI and WM databases.

For the MI data, we do the following: Since the MI database has training and testing datasets, we divide our classification system into a training-validation stage and a testing stage. For the training-validation stage, we first specify a cross-validation scheme of

10 iterations. For each iteration, 70% of the trials of the training dataset are randomly assigned to a training set and the remaining 30% to a validation set. Then, we use CKA (see Section 2.4) over the connectivity features obtained from the training set to generate a relevance vector $\varrho \in [0,1]^P$, where P equals the number of features in Φ. P varies according to the connectivity measure used to characterize the data. Then, we use ϱ to rank Φ. Next, we select a varying percentage of the ranked features, from 5% to 100% in 5% steps, and input them to the classification algorithm. The features associated with the highest values of ϱ are input first, and as the percentage of features increases those associated with lower values of ϱ are progressively included. In this work, we use a support vector classifier (SVC) with an RBF kernel [52]. All classification parameters, including the percentage of discriminant features, are tuned at this stage through a grid search. We select the parameters according to the classification accuracy, aiming to improve the system's performance. Then, for the testing stage, we train an SVC using the connectivity features from all trials in the training dataset as well as the parameters found in the previous stage. Lastly, we quantify the performance of the trained system in terms of the classification accuracy, obtained after predicting the MI task class labels of the testing dataset from its connectivity features.

The classification system we set up for the WM data closely resembles the one previously detailed for the MI data, with three changes. First, the WM database consists of one set of data for each subject, instead of two, so there is only a training-validation stage. Second, given the reduced number of trials available for each memory load level, each of the 10 iterations of the cross-validation scheme follows an 80–20% split for the training and validation sets (instead of a 70–30% split). Third, since the results provided by CKA are not stable for the low number of trials available from each subject (27.7 trials per class, on average), we opted to add an auxiliary cross-validation step, with the same characteristics as the one described above, and use it to estimate a single relevance vector $\bar{\varrho}$, obtained as the average of the relevance vectors of each data split. Then, we use $\bar{\varrho}$ to perform feature selection in every iteration of the main cross-validation scheme.

3.3. Parameter Selection

We used in-house Python implementations of the algorithms for all the connectivity measures studied (The $TE_{K\alpha}^{\theta}$ implementation is available at https://github.com/ide2704/Kernel_Phase_Transfer_Entropy, accessed on 13 July 2021), except for TE_{KSG}^{θ}. In that case, we used the implementation provided by the open access toolbox TRENTOOL, a TE estimation and analysis toolbox for Matlab [34].

Regarding the selection of parameters involved in the different effective connectivity estimation methods, we proceeded as follows: For the TE methods, we estimated all parameters from the real-valued time series, i.e., before extracting the phase time series. The embedding delay τ was set to 1 autocorrelation time (ACT), as proposed in [31]. The embedding dimension d was selected from the range $d = \{1, 2, \ldots, 10\}$ using Cao's criterion [34,53]. Note that for any signal pair, the embedding parameters selected are those of the driven or target time series, i.e., to estimate $TE(\mathbf{x} \rightarrow \mathbf{y})$ we use for both time series the embedding parameters found for $\mathbf{y}$. The interaction delay u was set as the value generating the largest TE from ranges that varied depending on the experiment: $u = \{1, 2, \ldots, 10\}$ for the NMMs, $u = \{1, 4, \ldots, 25\}$ for the MI data, and $u = \{50, 60, \ldots, 250\}$ for the WM data. Note that the meaning of u in terms of the time delay of the directed interaction between the driving and driven systems is associated with the sampling frequency, e.g., $u = \{1, 2, \ldots, 10\}$ for data sampled at 250 Hz translates to a time range between 4 ms and 40 ms. For $TE_{K\alpha}^{\theta}$ we select a value of $\alpha = 2$, which is neutral to weighting, a convenient choice when there is no previous knowledge about the values of the α parameter better suited for a particular application [10,28]. In addition, as kernel function, we employ an RBF kernel with Euclidean distance (see Equation (13)). The bandwidth σ was set in each case as the median distance of the data [54]. For TE_{KSG}^{θ} the Theiler correction window and the number of neighbors were left at their default values in TRENTOOL, 4 and 1 ACT, respectively [34]. For the GC methods the order of the autoregressive models o was selected from the range $o = \{1, 3, \ldots, 9\}$ using Akaike information criterion [55,56]. Furthermore, in order to estimate the PSI we employed a sliding window 5 frequency bins long (3 bins long for the WM data), centered on the frequency of interest. Finally, for all the connectivity methods involving the extraction of phase time series through Morlet wavelets, we varied the parameter m (see Equation (10)) from 3 to 10 in a logarithmic scale, according to the selected frequency of the filter.

4. Results and Discussion
4.1. Neural Mass Models Results

The experiments described in Section 3.1 are intended to assess whether the phase-based connectivity measures considered in this study correctly detect the direction of interaction between two time series of known oscillatory properties. Figure 5 presents the results obtained from such experiments. Namely, column **A** shows the connectivity values obtained for different levels of coupling strength, column **B** compares the connectivities estimated for ideal signals with those of signals contaminated with noise and mixing, and column **C** displays the results obtained for bidirectional narrowband couplings. The rows in Figure 5 correspond to each of the phase-based connectivity measures studied. The first row contains average PSI values computed on the frequency range between 2 Hz and 60 Hz, while rows two to five display average net connectivity values for $TE_{\kappa\alpha}^{\theta}$, TE_{KSG}^{θ}, TE_{Sym}^{θ}, and GC^{θ}, respectively. Circled values indicate statistically significant connectivities at a particular frequency, according to a permutation test based on randomized surrogate trials. The test identifies connectivity values that are, on average, significantly different from those expected for that connectivity measure applied to non-interacting signals. For the three experiments involving simulated data from NMMs, we use the PSI as a comparison standard, since it is a robust and well-stablished measure of linear directed interactions defined in terms of phase relations [12,13]. Therefore, it is suited to analyze the coupled, mildly nonlinear time series generated by NMMs.

Regarding the first experiment, which modifies the coupling strength between the simulated signals, the obtained results (Figure 5, column **A**) show that all the measures studied satisfactorily detect the coupling direction of the simulated data. Note that since we set the NMMs to generate unidirectional interactions from **x** to **y**, and because of the way we defined $\Delta\lambda$, then all net connectivity values for the simulated coupled signals should be positive. The same is true for the $PSI(x \to y)$. On the other hand, only the PSI, $TE_{\kappa\alpha}^{\theta}$, and GC^{θ} fulfill the criteria for an overall description of the phase-based interactions present in the data. First, we observe higher net connectivity values at higher coupling strengths, that is to say, stronger interactions lead to larger connectivity estimates. Second, for each coupling strength, there are higher net connectivity values around the frequencies corresponding to the main oscillatory components of the time series generated by the NMMs, in this case, oscillations in the range between 8 Hz and 20 Hz. Thirdly, there are statistically significant results for all the coupling strengths explored, except for non-interacting time series ($C_{12} = 0$). TE_{KSG}^{θ} does not capture statistically significant interactions for a coupling coefficient value of 0.2, indicating a lower sensitivity to weak couplings. While TE_{Sym}^{θ} exhibits a very distorted connectivity profile when compared with the PSI. In addition, it has much larger standard deviations for all the coupling strengths considered.

The second experiment assesses the robustness of our proposal to realistic levels of noise and signal mixing, two sources of signal degradation that can lead to spurious connectivity results. In electrophysiological signals, such as EEG, signal mixing arises as a result of field spread, while noise is the result of technical and physiological artifacts [9,57,58]. The results in Figure 5, column **B**, show that PSI, $TE_{\kappa\alpha}^{\theta}$, and GC^{θ} capture statistically significant interactions in the frequencies of interest for both the ideal (no noise or signal mixing) and realistic conditions. The smaller connectivity values for the data contaminated with noise and signal mixing, as compared with the ideal signals, are mostly explained by the reduction in asymmetry between the driving and driven signals caused by mixing [8]. On the contrary, we observe that neither TE_{Sym}^{θ} nor TE_{KSG}^{θ} produce statistically significant results under the realistic scenario, indicating that those estimators are less robust to signal degradation.

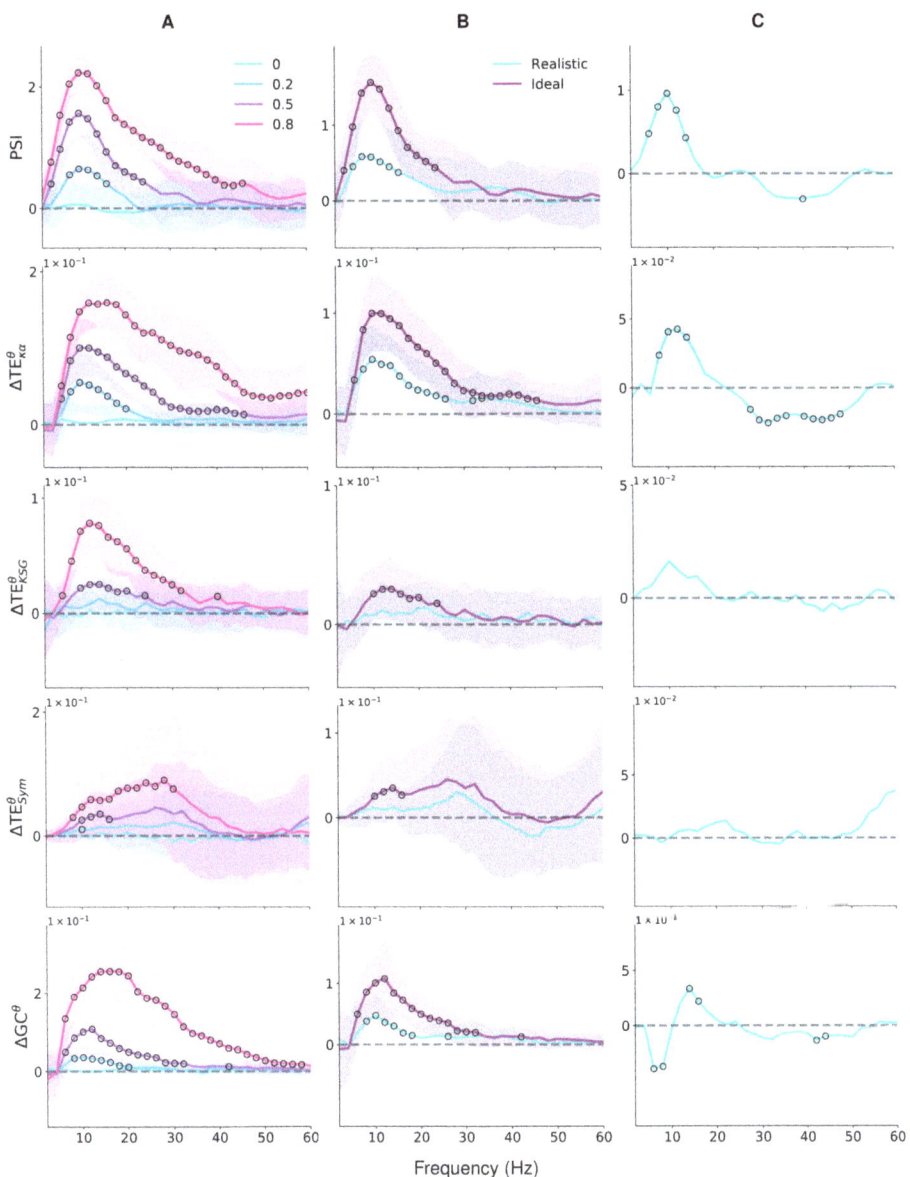

Figure 5. Obtained results for the experiments performed using simulated data from NMMs. Column (**A**) shows the average connectivity values obtained for different levels of coupling strength. Column (**B**) presents the average connectivity values estimated for ideal signals and for signals contaminated with noise and signal mixing. Column (**C**) displays the average connectivity values obtained for bidirectional narrowband couplings. The rows correspond to each of the net phase-based effective connectivity estimation approaches considered for the aforementioned experiments. Circled values indicate statistically significant results at a Bonferroni-corrected alpha level of 3.3×10^{-4}, according to a permutation test based on randomized surrogate trials.

Finally, the third experiment aims to evaluate how our proposal deals with bidirectional interactions of localized frequency content. Because of our experimental setup, the obtained results should exhibit a positive deflection around 10 Hz in order to capture the directed interaction from **x** to **y** and a negative deflection around 40 Hz to represent the directed interaction from **y** to **x**. Figure 5, column **C**, shows that both PSI and $TE^\theta_{\kappa\alpha}$ successfully detect the change in the direction of interaction in localized frequency bands, with statistically significant connectivity values around the frequencies of interest. However, under this scenario, $TE^\theta_{\kappa\alpha}$ is less frequency specific for high-frequency interactions than PSI, with statistically significant connections present for a large range of frequency values around 40 Hz. This is probably due to the filtering step involved in the estimation of $TE^\theta_{\kappa\alpha}$, while PSI is directly defined on the data spectra. Additionally, TE^θ_{KSG} and TE^θ_{Sym} fail to produce any significant results, while GC^θ shows a statistically significant, non-existing coupling from **y** to **x** for frequencies under 10 Hz. Note that, ultimately, the permutation test indicates whether the connectivity values obtained are unlikely to be the result of chance and not whether they correctly capture the directed interactions present in the data. In this case, the statistically significant results mean that GC^θ consistently found a directed interaction from **y** to **x** in the range mentioned before.

The results discussed above indicate that the proposed phase TE estimator is able to detect directed interactions between time series resembling electrophysiological data for different levels of coupling strength, under the presence of noise and signal mixing and for bidirectional narrowband couplings. Furthermore, they show that it is competitive with well-established approaches for phase-based net connectivity estimation, such as PSI, in the case of weakly nonlinear signals. Lastly, our results also show that commonly used single-trial TE estimators, such TE_{KSG} and TE_{Sym}, are ill-suited to measure directed interactions between instantaneous phase time series.

4.2. EEG Data Results

Table 1 presents the average accuracies achieved by the proposed classification systems for both the MI and WM databases, for each effective connectivity method studied. For the MI database, in the training-validation stage, the classifier based on $TE^\theta_{\kappa\alpha}$ features exhibited the highest average performance, closely followed by the one based on GC^θ. In the testing stage, we observe the same overall accuracy ranking, although a smaller drop in the classification accuracy occurs for $TE^\theta_{\kappa\alpha}$ than for GC^θ, which points to a better generalization capacity by the system trained using features extracted through phase TE. For the WM database, the classifier trained from $TE^\theta_{\kappa\alpha}$ features also displays the highest average accuracy. However, in this case, there is a large gap in performance between the $TE^\theta_{\kappa\alpha}$-based classification system and the closest results from an alternative approach. Furthermore, the results in Table 1 show a consistent improvement in performance between the classifiers that use real-valued TE estimates and those that are trained from phase TE values. They also show relatively low accuracies for the classifiers trained using PSI features. We believe the latter can be explained by two factors. First, by definition, the PSI is unable to explicitly detect bidirectional interactions. It measures connectivity in terms of lead/lag relations, which leads to ambiguity regarding the meaning of PSI values close to zero, since they can be the result of either the lack of interaction or evenly balanced bidirectional connections. If the relevant information to discriminate among the conditions of a cognitive paradigm is related to the bidirectionality of interactions, such as those present in WM [50,51], then the PSI might not be an adequate characterization strategy. Secondly, the PSI, like GC, is a linear measure; its performance degrades for strongly nonlinear phase relationships.

In the sections below, we detail and further discuss the results obtained for each database.

Table 1. MI and WM classification results in terms of the classification accuracy for all the effective connectivity measures considered.

	Motor Imagery (acc %)		Working Memory (acc %)
	Cross-Validation	Testing	Cross-Validation
GC [11]	64.3 ± 11.7	57.1 ± 11.0	53.0 ± 7.4
$TE_{\kappa\alpha}$ [10]	65.5 ± 11.4	62.8 ± 11.7	67.5 ± 4.2
PSI [12,51]	62.4 ± 7.8	58.8 ± 8.3	75.2 ± 5.2
GC^θ	67.0 ± 11.9	63.5 ± 14.4	74.5 ± 4.4
$TE_{\kappa\alpha}^\theta$	70.4 ± 12.5	69.0 ± 14.8	93.0 ± 5.9

4.2.1. Motor Imagery Results

Figure 6 depicts the average classification accuracy for all subjects in MI database as a function of the number of selected features during the training-validation stage, for $TE_{\kappa\alpha}$ and $TE_{\kappa\alpha}^\theta$. These results show there is a small improvement in the ability to discriminate between the MI tasks when using features extracted through phase TE, as compared with real-valued TE. In addition, they reveal that the CKA-based feature selection strategy successfully identified the most relevant connections for MI task classification. That is to say, the classification system has a stable performance even for a very reduced number of connectivity features. This is fundamental for any practical BCI application that intends to use phase TE as a characterization strategy, since estimating single-trial phase TE is computationally expensive [8]. Therefore, it is important to reduce as much as possible the number of channel pair connectivity features required to achieve peak classification performance. Additionally, it is important to highlight that while classification accuracies in Figure 6, and in Table 1, are in the same range of those obtained through other connectivity-based characterization approaches [10,23], they are far below those obtained from methods such as common spatial patterns [59–61]. A possible explanation is that bivariate TE might be more robust at describing long-range interactions rather than local ones [41], like those arising from MI-related activity, centered on the sensorimotor area. In addition, the differences with the results in [10], where we used $TE_{\kappa\alpha}$ to characterize the same database, lay mostly in the fact that in this study we select and analyze one 2 s long time window covering the period right after the end of the visual cue, while in [10] we report results from multiple overlapping windows covering the entirety of the task. Lastly, the large standard deviations from the average accuracies in Figure 6 point to disparate performances for different subjects.

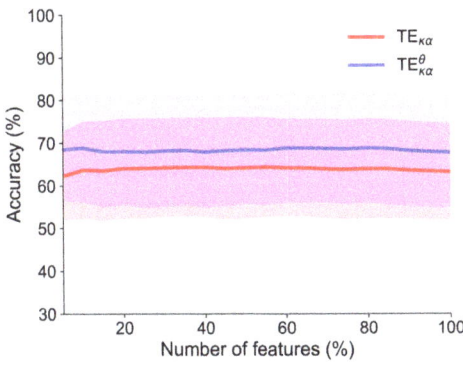

Figure 6. Average classification accuracies, and their standard deviations, for all subjects in the MI database as a function of the number features selected to train the classifiers.

Figure 7A shows the highest average classification accuracy per subject for $TE_{\kappa\alpha}^{\theta}$, GC^{θ} and PSI, during the training-validation stage. The subjects are ordered from highest to lowest performance. The analogous information for the testing stage is presented in Figure 7. In both stages, the $TE_{\kappa\alpha}^{\theta}$-based classifier performs slightly better than those based on alternative connectivity estimation strategies in most subjects. In addition, as inferred from Figure 6, there are large variations in performance for the different subjects in the database, consistent across the two classification stages. This behavior has been reported elsewhere [10,59–62].

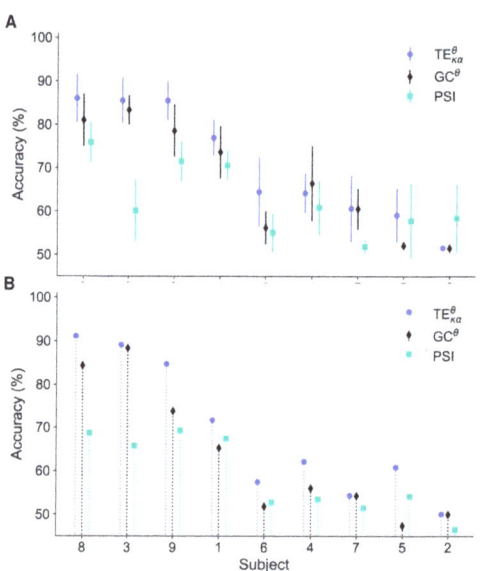

Figure 7. (**A**) Highest average classification accuracy for each subject in the MI database during the training-validation stage. (**B**) Accuracies obtained for each subject during the testing stage. The subjects are ordered from highest to lowest performance according to the accuracies obtained for the $TE_{\kappa\alpha}^{\theta}$-based classifier in the training-validation stage.

In order to gain insight into the observed performance differences, in the case of $TE_{\kappa\alpha}^{\theta}$, we exploited the second advantage provided by the CKA-based relevance analysis. The relevance vector index ϱ not only allows us to perform feature selection but also provides a one-to-one relevance mapping to each connectivity feature. That is to say, we can reconstruct normalized relevance connectivity matrices by properly reshaping ϱ, so as to visualize the connectivity pairs and frequency ranges that are discriminant for the task of interest. In that line, we followed the approach proposed in [23] to interpret the relevance information by clustering the subjects according to common relevance patterns.

First, for each subject and frequency band of interest, we obtained a relevance vector $\varrho_{n,\Delta f} \in \mathbb{R}^C$ whose elements were associated with each node (EEG channel) in the data by computing the relevance of the total information flow of every node. Such magnitude was defined as the sum of the relevance values ϱ, obtained from all data in the training dataset, corresponding to all directed interactions targeting and originating from a particular node. Then, we concatenated the vectors $\varrho_{n,\Delta f} \in \mathbb{R}^C$ for all frequency bands to obtain a single relevance vector $\varrho_n \in \mathbb{R}^{2C}$. Next, we reduced the dimension of the relevance vectors ϱ_n of each subject through t-Distributed Stochastic Neighbor Embedding (t-SNE), which preserves the spatial relationships existing in the initial higher-dimensional space [63]. Figure 8A shows the obtained two-dimensional representation of the relevance vectors for each subject in the MI database, colored according to their respective classification accuracy.

Note that the distribution of the subjects in the plot is related to their classification accuracies. This indicates that shared relevance patterns are related to the obtained classification results, meaning that subjects with similar ϱ_n had close performances. Then, we grouped the subjects into two clusters using the k-means algorithm. The number of clusters was selected by visual inspection of the t-SNE results. Figure 8B displays the two groups, termed G. I and G. II. The $\text{TE}_{\kappa\alpha}^{\theta}$-based classifier has average accuracies of 0.59 ± 0.05 and 0.80 ± 0.09 for the subjects in G. I and G. II, respectively.

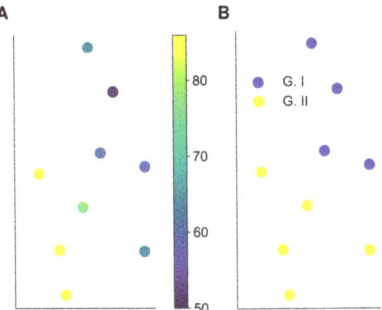

Figure 8. (**A**) Two-dimensional representation of the relevance vectors for each subject in the MI database obtained after applying t-SNE on ϱ_n. (**B**) Groups identified by k-means. For the $\text{TE}_{\kappa\alpha}^{\theta}$-based classifier the subjects grouped in G. I have an average accuracy of 0.59 ± 0.05, while those in G. II have an average accuracy of 0.80 ± 0.09.

Finally, Figure 9 shows the average nodal relevance, as defined by ϱ_n, and the most relevant connectivities for each group, discriminated by frequency band. For G. I we observe high node relevance mostly in the α band in right fronto-central, left-central, and centro-parietal regions. The most relevant connections in the α band tend to originate or target fronto-central nodes, while the ones in the β_l band favor parietal and centro-parietal areas. For G. II, the node relevance is concentrated around the right centro-parietal region, particularly channel CP4, for both frequency bands. The most relevant connections in the α band involve short-range interactions mainly between centro-parietal and central regions. The most relevant connections in the β_l band, which display higher values than those of α, originate from CP3 and CP4 and target central and fronto-central nodes. Since the G. II includes all the subjects with good classification performances, we can conclude that the information that allows to satisfactorily classify the left and right hand MI tasks from $\text{TE}_{\kappa\alpha}^{\theta}$ features corresponds mostly to the incoming and outgoing information flow coded in the phases of the oscillatory activity in the centro-parietal region. These results are in line, in terms of spatial location, with those we found in [10], and with physiological interpretations that argue that MI activates motor representations in the parietal areal and the premotor cortex [64].

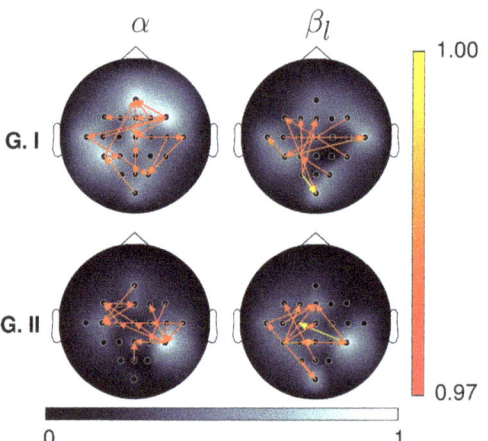

Figure 9. Topoplots of the average node (channel) relevance for each group of clustered subjects and frequency band of interest in the MI database (see Figure 8). The arrows represent the most relevant connectivities for each group. For visualization purposes, only 3% of the connections, those with the highest average relevance values per group, are depicted.

4.2.2. Working Memory Results

Figure 10 presents the average classification accuracy for all subjects in the WM database as a function of the number of selected features, for $TE_{\kappa\alpha}$ and $TE_{\kappa\alpha}^{\theta}$. The results show that the classifier trained from phase TE features markedly outperforms the one trained using real-valued TE estimates, as long as the appropriate percentage of features is selected. This difference might be attributed to the hypothesized phase-based nature of directed interactions during WM tasks [35,50], which would be better captured by phase TE. Furthermore, both accuracy curves highlight the importance of feature selection, since they show a steep performance degradation as more features are used to train the classifiers. In this case, the CKA-based relevance analysis not only allows reducing the number of features needed to successfully classify the three cognitive load levels present in the WM data but also prevents the classifiers from being confounded by connections that do not hold relevant information to discriminate between the target conditions.

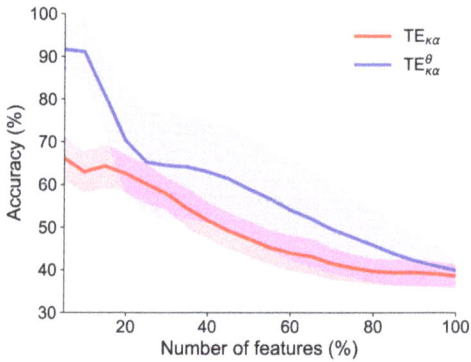

Figure 10. Average classification accuracies, and their standard deviations, for all subjects in the WM database as a function of the number features selected to train the classifiers.

Figure 11 depicts the highest average classification accuracy per subject for $TE_{\kappa\alpha}^{\theta}$, GC^{θ} and PSI. The subjects are ordered from highest to lowest performance. Unlike the results obtained for the MI database, we do not observe an underperforming group of subjects, especially after considering the fact that for the WM database the classifiers must discriminate among three classes instead of two. On the other hand, in this case, the $TE_{\kappa\alpha}^{\theta}$-based classifier largely outperforms those based on alternative connectivity estimation strategies in most subjects. Here, we must point out that the auxiliary cross-validation step introduced for feature selection, aiming to obtain stable CKA results for the reduced number of available trials, leads to data leakage. This is because, ultimately, it requires all the available data to estimate $\bar{\varrho}$, which renders it a nonviable approach for practical BCI implementations and can inflate performance evaluations, such as the accuracy results previously discussed. However, since the same strategy was implemented for all classification systems and connectivity measures considered for the WM database, comparisons among them remain valid, and the relative differences in performance are still informative.

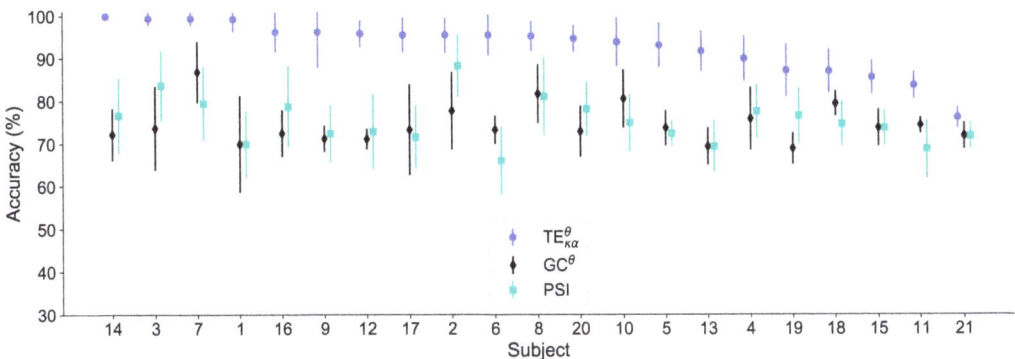

Figure 11. Highest average classification accuracy for each subject in the WM database. The subjects are ordered from highest to lowest performance according to the accuracies obtained for the $TE_{\kappa\alpha}^{\theta}$-based classifier.

In order to elucidate the pairwise connectivities, and their corresponding frequency bands, that allow the $TE_{\kappa\alpha}^{\theta}$-based classification system to successfully discriminate among different memory loads, we proceeded as described in Section 4.2.1 and from $\bar{\varrho}$ obtained a node relevance vector $\bar{\varrho}_n \in \mathbb{R}^{3C}$. Then, we applied t-SNE on $\bar{\varrho}_n$. Figure 12A shows the obtained two-dimensional representation of the relevance vectors for each subject in the WM database. Unlike the results observed before for the MI database, there is not a clear association between the subject distribution on the plot and their classification accuracies. Nonetheless, Figure 12A shows the presence of well-defined groups sharing similar relevance patterns. As before, we grouped the subjects into clusters using the k-means algorithm. The number of clusters was selected as three by visual inspection of the t-SNE results. Figure 12B displays the three groups, termed G. I, G. II, and G. III. The $TE_{\kappa\alpha}^{\theta}$-based classifier has average accuracies of 0.94 ± 0.04, 0.92 ± 0.08, and 0.93 ± 0.08 for the subjects in G. I, G. II, and G. III, respectively.

Lastly, Figure 13 shows the average nodal relevance, as defined by ϱ_n, and the most relevant connectivities for each group, discriminated by frequency band. For G. I we observe widespread high node relevance in both the α and β_l bands and low node relevance in the θ band. Most relevant connections are present in the β_l band with many connections originating in the parieto-occipital region and targeting frontal and centro-frontal areas. For G. II and G. III node relevance is more evenly distributed across the three frequency bands considered. Spatially, it is more prominent around some pre-frontal, frontal, centro-parietal, and parietal nodes. In terms of the most relevant connections, we observe long-range contralateral interactions involving mostly the regions previously listed, as well as some connections to and from temporal areas. Therefore, we argue that the information flow between frontal, parietal, and temporal regions, coded in the phases of oscillatory activity in the θ, α, and β_l bands, is what allowed us to discriminate among different memory loads from $\text{TE}_{\kappa\alpha}^{\theta}$ features. These results agree with several studies that identify fronto-parietal and fronto-temporal neural circuits operating in frequency ranges spanning from θ to β as key during the activation of working memory [35,50,51].

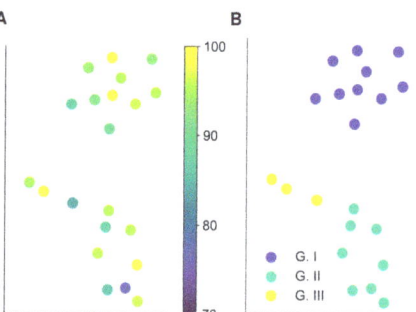

Figure 12. (**A**) Two-dimensional representation of the relevance vectors for each subject in the WM database obtained after applying t-SNE on ϱ_n. (**B**) Groups identified by k-means. For the $\text{TE}_{\kappa\alpha}^{\theta}$-based classifier the subjects grouped in G. I, have an average accuracy of 0.94 ± 0.04, while those in G. II and G.III have average accuracies of 0.92 ± 0.08 and 0.93 ± 0.08, respectively.

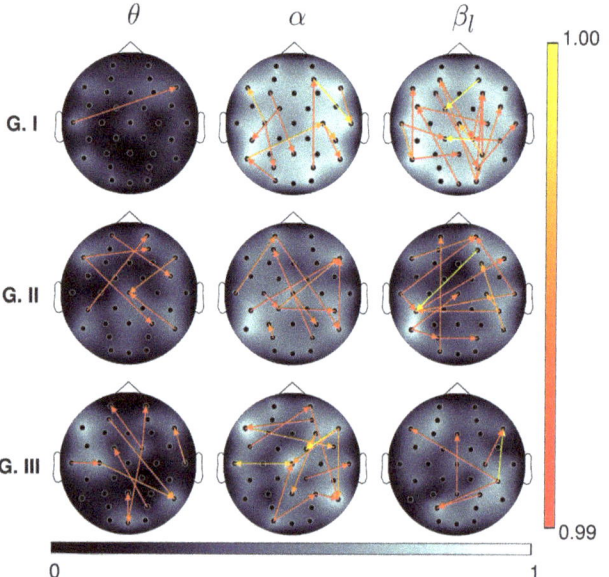

Figure 13. Topoplots of the average node (channel) relevance for each group of clustered subjects and frequency band of interest in the WM database (see Figure 12). The arrows represent the most relevant connectivities for each group. For visualization purposes, only the 1% of the connections, those with the highest average relevance values per group, are depicted.

4.3. Limitations

In this study, we employed Morlet wavelets as filters for instantaneous phase extraction prior to phase TE estimation, as proposed in [8]. However, as discussed by the authors in [8], the choice of filter can influence the behavior of phase TE. This is an aspect we have yet to explore for our proposal. In the same line, in [42] the authors showed, using the Kraskov-Stögbauer-Grassberger TE estimator on real-valued filtered signals, that filtering and downsampling are deleterious for TE, since they can lead to altered time delays and hide certain causal interactions. Furthermore, from a conceptual perspective, while filtering dampens spectral power, it does not always remove the information contained in specific frequencies [25]. This would hinder the isolation of frequency specific interactions in TE estimates from real-valued filtered data, the most common approach to obtain spectrally resolved TE values. Whether those effects are also present in the case of phase TE is yet to be analyzed; however, as pointed out in [25], phase TE is conceptually different from spectrally resolved TE. Additionally, the results obtained with our phase TE estimator for the NMM data closely follow those obtained with the PSI, a measure that does not rely on data filtering, which points to a certain degree of robustness to the negative effects that might be associated with phase extraction through complex filtering. A related issue is the possible effects on our proposal of the preprocessing pipelines employed on the EEG data, which involve spectral and spatial filtering. Regarding the former, we have not studied its effects in this work; while for the latter, surface Laplacian positively impacted the discrimination capability of the connectivity features obtained from the different measures considered.

In addition, we are yet to examine the effects of the parameter α in Renyi's entropy on the proposed phase TE estimator. In [10], we showed that the choice of α indeed modified the performance of the $TE_{\kappa\alpha}$. The same must hold true for $TE_{\kappa\alpha}^{\theta}$. Additionally, we selected the autocorrelation time and Cao's criterion to obtain the embedding parameters for all the TE estimation methods. More complex approaches such as time-delayed mutual information and Ragwitz criterion may yield better results [34]. However, since our motivation was to propose a single-trial phase TE estimator suited as characterization method for BCI applications, the choice of simple parameter estimation methods is justified. As a matter of fact, a practical implementation of a phase TE-based BCI system would likely require further simplifications regarding parameter estimation, in order to facilitate the computation of phase TE in real time. Furthermore, our proposed phase TE estimator inherits the limitations of $TE_{\kappa\alpha}$ [10]. Namely, it is ill suited to analyze long time series (several thousands of data points) because of the increase in computational cost, especially for non-integer values of the parameter α. In addition, it assumes stationary or weakly non-stationary data. Finally, since the definition of causality underlying TE is observational, the proposed phase TE estimator is blind to unobserved common causes, including those resulting from different driving delays.

5. Conclusions

In this work, we proposed a single-trial phase TE estimator. Our method combines a kernel-based TE estimation approach, which defines effectivity connectivity as a linear combination of Renyi's entropy measures of order α, with instantaneous phase time series extracted from the data under analysis. We tested the performance of our proposal on synthetic data generated through NMMs and on two EEG databases obtained under MI and WM paradigms. We compared it with commonly used single-trial TE estimators, applied to phase time series, and the PSI and GC. Our results show that the proposed phase TE estimator successfully detects the direction of interaction between individual pairs of signals, capturing the differences in coupling strength and displaying statistically significant results around the frequencies corresponding to the main oscillatory components present in the data. It also succeeds in detecting bidirectional interactions of localized frequency content and is robust to realistic noise and signal mixing levels. Moreover, our method, coupled with a CKA-based relevance analysis, revealed discriminant spatial and frequency-dependent patterns for both the MI and WM databases, leading to improved classification performance compared with approaches based on real-valued TE estimation. In all our experiments, the proposed single-trial kernel-based phase TE estimator is competitive with the comparison methods previously listed in terms of the performance assessment metrics employed.

As future work, we will look into developing a cross-spectral representation for our phase TE estimator to study directed interactions between oscillations of different frequencies [65]. We will also explore the effects of the choice of filter on the proposed estimator as well as those of the parameters involved in time embedding and in our kernel-based TE estimation approach.

Author Contributions: Conceptualization, I.D.L.P.P. and A.Á.-M.; methodology, I.D.L.P.P., A.Á.-M. and P.M.H.G.; software, I.D.L.P.P., A.Á.-M. and D.C.-P.; validation, I.D.L.P.P., A.Á.-M. and D.C.-P.; formal analysis, I.D.L.P.P. and Á.O.-G.; investigation, I.D.L.P.P., A.Á.-M. and J.I.R.P.; resources, I.D.L.P.P., A.Á.-M. and Á.O.-G.; data curation, I.D.L.P.P. and P.M.H.G.; writing—original draft preparation, I.D.L.P.P. and A.Á.-M.; writing—review and editing, A.Á.-M., P.M.H.G. and D.C.-P.; visualization, I.D.L.P.P.; supervision, Á.O.-G. and A.Á.-M.; project administration, Á.O.-G.; funding acquisition, I.D.L.P.P., A.Á.-M. and J.I.R.P. All authors have read and agreed to the published version of the manuscript.

Funding: Under grants provided by: The Minciencias project (111080763051)-Herramienta de apoyo al diagnóstico del TDAH en niños a partir de múltiples características de actividad cerebral desde registros EEG; Maestría en ingeniería eléctrica and Maestría en Ingeniería de Sistemas y Computación—Universidad Tecnológica de Pereira. Author Iván De La Pava Panche was supported by the program "Doctorado Nacional en Empresa-Convoctoria 758 de 2016", funded by Minciencias.

Institutional Review Board Statement: In this study, we use public-access EEG databases introduced in previously published works and made freely available by the respective authors [44,48]. We did not collect any data from human participants ourselves.

Informed Consent Statement: This study uses anonymized public databases introduced in previously published works by other groups [44,48].

Data Availability Statement: The databases used in this study are public and can be found at the following links: BCI Competition IV database 2a http://www.bbci.de/competition/iv/index.html (accessed on 2 June 2021), database from brain activity during visual working memory https://data.mendeley.com/datasets/j2v7btchdy/2 (accessed on 2 June 2021).

Conflicts of Interest: The authors declare that this research was conducted in the absence of any commercial or financial relationships that could be construed as a potential conflict of interest.

References

1. La Tour, T.D.; Tallot, L.; Grabot, L.; Doyère, V.; Van Wassenhove, V.; Grenier, Y.; Gramfort, A. Non-linear auto-regressive models for cross-frequency coupling in neural time series. *PLoS Comput. Biol.* **2017**, *13*, e1005893.
2. Da Silva, F.L. EEG: Origin and measurement. In *EEg-fMRI*; Springer: Berlin/Heidelberg, Germany, 2009; pp. 19–38.
3. Wianda, E.; Ross, B. The roles of alpha oscillation in working memory retention. *Brain Behav.* **2019**, *9*, e01263. [CrossRef] [PubMed]
4. Hyafil, A.; Giraud, A.L.; Fontolan, L.; Gutkin, B. Neural cross-frequency coupling: Connecting architectures, mechanisms, and functions. *Trends Neurosci.* **2015**, *38*, 725–740. [CrossRef] [PubMed]
5. Xie, P.; Pang, X.; Cheng, S.; Zhang, Y.; Yang, Y.; Li, X.; Chen, X. Cross-frequency and iso-frequency estimation of functional corticomuscular coupling after stroke. *Cogn. Neurodyn.* **2021**, *15*, 439–451. [CrossRef] [PubMed]
6. Ahmadi, A.; Davoudi, S.; Behroozi, M.; Daliri, M.R. Decoding covert visual attention based on phase transfer entropy. *Physiol. Behav.* **2020**, *222*, 112932. [CrossRef]
7. Kang, H.; Zhang, X.; Zhang, G. Phase permutation entropy: A complexity measure for nonlinear time series incorporating phase information. *Phys. A Stat. Mech. Appl.* **2021**, *568*, 125686. [CrossRef]
8. Lobier, M.; Siebenhühner, F.; Palva, S.; Palva, J.M. Phase transfer entropy: A novel phase-based measure for directed connectivity in networks coupled by oscillatory interactions. *Neuroimage* **2014**, *85*, 853–872. [CrossRef]
9. Sakkalis, V. Review of advanced techniques for the estimation of brain connectivity measured with EEG/MEG. *Comput. Biol. Med.* **2011**, *41*, 1110–1117. [CrossRef]
10. De La Pava Panche, I.; Alvarez-Meza, A.M.; Orozco-Gutierrez, A. A data-driven measure of effective connectivity based on Renyi's α-entropy. *Front. Neurosci.* **2019**, *13*, 1277. [CrossRef] [PubMed]
11. Cekic, S.; Grandjean, D.; Renaud, O. Time, frequency, and time-varying Granger-causality measures in neuroscience. *Stat. Med.* **2018**, *37*, 1910–1931. [CrossRef]
12. Nolte, G.; Ziehe, A.; Nikulin, V.V.; Schlögl, A.; Krämer, N.; Brismar, T.; Müller, K.R. Robustly estimating the flow direction of information in complex physical systems. *Phys. Rev. Lett.* **2008**, *100*, 234101. [CrossRef] [PubMed]
13. Jiang, H.; Bahramisharif, A.; van Gerven, M.A.; Jensen, O. Measuring directionality between neuronal oscillations of different frequencies. *Neuroimage* **2015**, *118*, 359–367. [CrossRef]
14. Schreiber, T. Measuring information transfer. *Phys. Rev. Lett.* **2000**, *85*, 461. [CrossRef] [PubMed]

15. Zhu, J.; Bellanger, J.J.; Shu, H.; Le Bouquin Jeannès, R. Contribution to transfer entropy estimation via the k-nearest-neighbors approach. *Entropy* **2015**, *17*, 4173–4201. [CrossRef]
16. Wilmer, A.; de Lussanet, M.; Lappe, M. Time-delayed mutual information of the phase as a measure of functional connectivity. *PLoS ONE* **2012**, *7*, e44633. [CrossRef]
17. Numan, T.; Slooter, A.J.; van der Kooi, A.W.; Hoekman, A.M.; Suyker, W.J.; Stam, C.J.; van Dellen, E. Functional connectivity and network analysis during hypoactive delirium and recovery from anesthesia. *Clin. Neurophysiol.* **2017**, *128*, 914–924. [CrossRef]
18. Hillebrand, A.; Tewarie, P.; Van Dellen, E.; Yu, M.; Carbo, E.W.; Douw, L.; Gouw, A.A.; Van Straaten, E.C.; Stam, C.J. Direction of information flow in large-scale resting-state networks is frequency-dependent. *Proc. Natl. Acad. Sci. USA* **2016**, *113*, 3867–3872. [CrossRef] [PubMed]
19. Wang, S.; Zhang, D.; Fang, B.; Liu, X.; Yan, G.; Sui, G.; Huang, Q.; Sun, L.; Wang, S. A Study on Resting EEG Effective Connectivity Difference before and after Neurofeedback for Children with ADHD. *Neuroscience* **2021**, *457*, 103–113. [CrossRef]
20. Yang, P.; Shang, P.; Lin, A. Financial time series analysis based on effective phase transfer entropy. *Phys. A Stat. Mech. Appl.* **2017**, *468*, 398–408. [CrossRef]
21. Rathee, D.; Cecotti, H.; Prasad, G. Single-trial effective brain connectivity patterns enhance discriminability of mental imagery tasks. *J. Neural Eng.* **2017**, *14*, 056005. [CrossRef]
22. Zhang, R.; Li, X.; Wang, Y.; Liu, B.; Shi, L.; Chen, M.; Zhang, L.; Hu, Y. Using brain network features to increase the classification accuracy of MI-BCI inefficiency subject. *IEEE Access* **2019**, *7*, 74490–74499. [CrossRef]
23. García-Murillo, D.G.; Alvarez-Meza, A.; Castellanos-Dominguez, G. Single-Trial Kernel-Based Functional Connectivity for Enhanced Feature Extraction in Motor-Related Tasks. *Sensors* **2021**, *21*, 2750. [CrossRef] [PubMed]
24. Chen, X.; Zhang, Y.; Cheng, S.; Xie, P. Transfer spectral entropy and application to functional corticomuscular coupling. *IEEE Trans. Neural Syst. Rehabil. Eng.* **2019**, *27*, 1092–1102. [CrossRef] [PubMed]
25. Pinzuti, E.; Wollstadt, P.; Gutknecht, A.; Tüscher, O.; Wibral, M. Measuring spectrally-resolved information transfer. *PLoS Comput. Biol.* **2020**, *16*, e1008526. [CrossRef] [PubMed]
26. Rényi, A. On measures of entropy and information. In *Proceedings of the Fourth Berkeley Symposium on Mathematical Statistics and Probability, Volume 1: Contributions to the Theory of Statistics*; The Regents of the University of California: California, CA, USA, 1961.
27. Principe, J.C. *Information Theoretic Learning: Renyi's Entropy and Kernel Perspectives*; Springer Science & Business Media: New York, NY, USA, 2010.
28. Giraldo, L.G.S.; Rao, M.; Principe, J.C. Measures of entropy from data using infinitely divisible kernels. *IEEE Trans. Inf. Theory* **2015**, *61*, 535–548. [CrossRef]
29. Cortes, C.; Mohri, M.; Rostamizadeh, A. Algorithms for learning kernels based on centered alignment. *J. Mach. Learn. Res.* **2012**, *13*, 795–828.
30. Wibral, M.; Pampu, N.; Priesemann, V.; Siebenhühner, F.; Seiwert, H.; Lindner, M.; Lizier, J.T.; Vicente, R. Measuring information-transfer delays. *PLoS ONE* **2013**, *8*, e55809. [CrossRef] [PubMed]
31. Vicente, R.; Wibral, M.; Lindner, M.; Pipa, G. Transfer entropy—A model-free measure of effective connectivity for the neurosciences. *J. Comput. Neurosci.* **2011**, *30*, 45–67. [CrossRef] [PubMed]
32. Takens, F. Detecting strange attractors in turbulence. In *Dynamical Systems and Turbulence, Warwick 1980*; Springer: Berlin/Heidelberg, Germany, 1981; pp. 366–381.
33. Kraskov, A.; Stögbauer, H.; Grassberger, P. Estimating mutual information. *Phys. Rev. E* **2004**, *69*, 066138. [CrossRef]
34. Lindner, M.; Vicente, R.; Priesemann, V.; Wibral, M. TRENTOOL: A Matlab open source toolbox to analyse information flow in time series data with transfer entropy. *BMC Neurosci.* **2011**, *12*, 119. [CrossRef]
35. Dimitriadis, S.; Sun, Y.; Laskaris, N.; Thakor, N.; Bezerianos, A. Revealing cross-frequency causal interactions during a mental arithmetic task through symbolic transfer entropy: A novel vector-quantization approach. *IEEE Trans. Neural Syst. Rehabil. Eng.* **2016**, *24*, 1017–1028. [CrossRef]
36. Barnett, L.; Barrett, A.B.; Seth, A.K. Granger causality and transfer entropy are equivalent for Gaussian variables. *Phys. Rev. Lett.* **2009**, *103*, 238701. [CrossRef]
37. Liu, W.; Principe, J.C.; Haykin, S. *Kernel Adaptive Filtering: A Comprehensive Introduction*; John Wiley & Sons: Hoboken, NJ, USA, 2011; Volume 57.
38. Fernández-Ramírez, J.; Álvarez-Meza, A.; Pereira, E.; Orozco-Gutiérrez, A.; Castellanos-Dominguez, G. Video-based social behavior recognition based on kernel relevance analysis. *Vis. Comput.* **2020**, *36*, 1535–1547. [CrossRef]
39. David, O.; Friston, K.J. A neural mass model for MEG/EEG:: Coupling and neuronal dynamics. *NeuroImage* **2003**, *20*, 1743–1755. [CrossRef]
40. David, O.; Cosmelli, D.; Friston, K.J. Evaluation of different measures of functional connectivity using a neural mass model. *Neuroimage* **2004**, *21*, 659–673. [CrossRef] [PubMed]
41. Ursino, M.; Ricci, G.; Magosso, E. Transfer Entropy as a Measure of Brain Connectivity: A Critical Analysis With the Help of Neural Mass Models. *Front. Comput. Neurosci.* **2020**, *14*, 45. [CrossRef]
42. Weber, I.; Florin, E.; Von Papen, M.; Timmermann, L. The influence of filtering and downsampling on the estimation of transfer entropy. *PLoS ONE* **2017**, *12*, e0188210. [CrossRef]

43. Collazos-Huertas, D.; Álvarez-Meza, A.; Acosta-Medina, C.; Castaño-Duque, G.; Castellanos-Dominguez, G. CNN-based framework using spatial dropping for enhanced interpretation of neural activity in motor imagery classification. *Brain Inform.* **2020**, *7*, 1–13. [CrossRef] [PubMed]
44. Tangermann, M.; Müller, K.R.; Aertsen, A.; Birbaumer, N.; Braun, C.; Brunner, C.; Leeb, R.; Mehring, C.; Miller, K.J.; Mueller-Putz, G.; et al. Review of the BCI competition IV. *Front. Neurosci.* **2012**, *6*, 55.
45. Perrin, F.; Pernier, J.; Bertrand, O.; Echallier, J. Spherical splines for scalp potential and current density mapping. *Electroencephalogr. Clin. Neurophysiol.* **1989**, *72*, 184–187. [CrossRef]
46. Cohen, M.X. Comparison of different spatial transformations applied to EEG data: A case study of error processing. *Int. J. Psychophysiol.* **2015**, *97*, 245–257. [CrossRef] [PubMed]
47. Zhang, D.; Zhao, H.; Bai, W.; Tian, X. Functional connectivity among multi-channel EEGs when working memory load reaches the capacity. *Brain Res.* **2016**, *1631*, 101–112. [CrossRef] [PubMed]
48. Villena-González, M.; Rubio-Venegas, I.; López, V. Data from brain activity during visual working memory replicates the correlation between contralateral delay activity and memory capacity. *Data Brief* **2020**, *28*, 105042. [CrossRef] [PubMed]
49. Vogel, E.K.; Machizawa, M.G. Neural activity predicts individual differences in visual working memory capacity. *Nature* **2004**, *428*, 748–751. [CrossRef] [PubMed]
50. Johnson, E.L.; Adams, J.N.; Solbakk, A.K.; Endestad, T.; Larsson, P.G.; Ivanovic, J.; Meling, T.R.; Lin, J.J.; Knight, R.T. Dynamic frontotemporal systems process space and time in working memory. *PLoS Biol.* **2018**, *16*, e2004274. [CrossRef]
51. Johnson, E.L.; King-Stephens, D.; Weber, P.B.; Laxer, K.D.; Lin, J.J.; Knight, R.T. Spectral imprints of working memory for everyday associations in the frontoparietal network. *Front. Syst. Neurosci.* **2019**, *12*, 65. [CrossRef] [PubMed]
52. Pedregosa, F.; Varoquaux, G.; Gramfort, A.; Michel, V.; Thirion, B.; Grisel, O.; Blondel, M.; Prettenhofer, P.; Weiss, R.; Dubourg, V.; et al. Scikit-learn: Machine Learning in Python. *J. Mach. Learn. Res.* **2011**, *12*, 2825–2830.
53. Cao, L. Practical method for determining the minimum embedding dimension of a scalar time series. *Phys. D Nonlinear Phenom.* **1997**, *110*, 43–50. [CrossRef]
54. Schölkopf, B.; Smola, A.J.; Bach, F. *Learning with Kernels: Support Vector Machines, Regularization, Optimization, and Beyond*; MIT Press: Cambridge, MA, USA, 2002.
55. Akaike, H. A new look at the statistical model identification. *IEEE Trans. Autom. Control* **1974**, *19*, 716–723. [CrossRef]
56. Gong, A.; Liu, J.; Chen, S.; Fu, Y. Time–Frequency Cross Mutual Information Analysis of the Brain Functional Networks Underlying Multiclass Motor Imagery. *J. Mot. Behav.* **2018**, *50*, 254–267. [CrossRef]
57. Debener, S.; Minow, F.; Emkes, R.; Gandras, K.; De Vos, M. How about taking a low-cost, small, and wireless EEG for a walk? *Psychophysiology* **2012**, *49*, 1617–1621. [CrossRef]
58. Mennes, M.; Wouters, H.; Vanrumste, B.; Lagae, L.; Stiers, P. Validation of ICA as a tool to remove eye movement artifacts from EEG/ERP. *Psychophysiology* **2010**, *47*, 1142–1150. [CrossRef]
59. Li, D.; Zhang, H.; Khan, M.S.; Mi, F. A self-adaptive frequency selection common spatial pattern and least squares twin support vector machine for motor imagery electroencephalography recognition. *Biomed. Signal Process. Control* **2018**, *41*, 222–232. [CrossRef]
60. Gómez, V.; Álvarez, A.; Herrera, P.; Castellanos, G.; Orozco, A. Short Time EEG Connectivity Features to Support Interpretability of MI Discrimination. In *Iberoamerican Congress on Pattern Recognition*; Springer: Cham, Switzerland, 2018; pp. 699–706.
61. Elasuty, B.; Eldawlatly, S. Dynamic Bayesian Networks for EEG motor imagery feature extraction. In Proceedings of the 2015 7th International IEEE/EMBS Conference on Neural Engineering (NER), Montpellier, France, 22–24 April 2015; pp. 170–173.
62. Liang, S.; Choi, K.S.; Qin, J.; Wang, Q.; Pang, W.M.; Heng, P.A. Discrimination of motor imagery tasks via information flow pattern of brain connectivity. *Technol. Health Care* **2016**, *24*, S795–S801. [CrossRef] [PubMed]
63. Linderman, G.C.; Steinerberger, S. Clustering with t-SNE, provably. *SIAM J. Math. Data Sci.* **2019**, *1*, 313–332. [CrossRef]
64. Hétu, S.; Grégoire, M.; Saimpont, A.; Coll, M.P.; Eugène, F.; Michon, P.E.; Jackson, P.L. The neural network of motor imagery: An ALE meta-analysis. *Neurosci. Biobehav. Rev.* **2013**, *37*, 930–949. [CrossRef] [PubMed]
65. Martínez-Cancino, R.; Delorme, A.; Wagner, J.; Kreutz-Delgado, K.; Sotero, R.C.; Makeig, S. What can local transfer entropy tell us about phase-amplitude coupling in electrophysiological signals? *Entropy* **2020**, *22*, 1262. [CrossRef] [PubMed]

Article

Monitoring Brain State and Behavioral Performance during Repetitive Visual Stimulation

Alexander K. Kuc [1,*,†], Semen A. Kurkin [1,2,†], Vladimir A. Maksimenko [1,2,3,4,†], Alexander N. Pisarchik [2,5] and Alexander E. Hramov [1,2,3,4]

1. Baltic Center for Artificial Intelligence and Neurotechnology, Immanuel Kant Baltic Federal University, 236040 Kaliningrad, Russia; s.kurkin@innopolis.ru (S.A.K.); v.maksimenko@innopolis.ru (V.A.M.); a.hramov@innopolis.ru (A.E.H.)
2. Center for Technologies in Robotics and Mechatronics Components, Innopolis University, 420500 Innopolis, Russia; alexander.pisarchik@ctb.upm.es
3. Neurotechnology Deparment, Lobachevsky State University of Nizhny Novgorod, 603022 Nizhny Novgorod, Russia
4. Department of Innovative Cardiological Information Technology, Institute of Cardiological Research, Saratov State Medical University, 410012 Saratov, Russia
5. Centre for Biomedical Technology, Universidad Piolitécnica de Madrid, Pozuelo de Alarcón, 28223 Madrid, Spain
* Correspondence: plo@sstu.ru
† These authors contributed equally to this work.

Abstract: We tested whether changes in prestimulus neural activity predict behavioral performance (decision time and errors) during a prolonged visual task. The task was to classify ambiguous stimuli—Necker cubes; manipulating the degree of ambiguity from low ambiguity (LA) to high ambiguity (HA) changed the task difficulty. First, we assumed that the observer's state changes over time, which leads to a change in the prestimulus brain activity. Second, we supposed that the prestimulus state produces a different effect on behavioral performance depending on the task demands. Monitoring behavioral responses, we revealed that the observer's decision time decreased for both LA and HA stimuli during the task performance. The number of perceptual errors lowered for HA, but not for LA stimuli. EEG analysis revealed an increase in the prestimulus 9–11 Hz EEG power with task time. Finally, we found associations between the behavioral and neural estimates. The prestimulus EEG power negatively correlated with the decision time for LA stimuli and the erroneous responses rate for HA stimuli. The obtained results confirm that monitoring prestimulus EEG power enables predicting perceptual performance on the behavioral level. The observed different time-on-task effects on the LA and HA stimuli processing may shed light on the features of ambiguous perception.

Keywords: ambiguous stimuli; Necker cubes; classification task; EEG analysis; wavelet analysis; decision time; perceptual errors; time-on-task effect

1. Introduction

Sensory processing is a fundamental brain function that allows us to more easily interact with each other and with our environment. In everyday life, we collect sensory data and process it for interpretation and decision making [1]. The accuracy and timeliness of our decisions depend on the speed and correctness of sensory processing. The effectiveness of sensory processing, in turn, is determined by a number of exogenous and endogenous factors [2]. In particular, the exogenous component reflects the quality of the sensory input. Thus, when faced with unambiguous information, we can easily interpret it. On the contrary, when information becomes ambiguous, interpreting it takes more effort.

In turn, the endogenous component depends on the state of the person; on their attention, fatigue, and subjective experience [3]. In many experimental studies where ambiguous stimuli were used, endogenous effects were found to be especially pronounced when the sensory information quality was low [4]. Therefore, the observer must concentrate to gather more information to make the right decision, relying on personal experience to extrapolate limited information or unresolve its ambiguity.

The conditions under which the observer receives and processes information are important as well. For example, high-speed driving on a rainy night requires the very fast processing of low-quality information. Performing monotonous tasks with increased responsibility (e.g., flight or power plant operators) also requires maintaining high performance and emergence preparedness. In these stressful conditions, the influence of exogenous and endogenous factors on the likelihood of perceptual errors should be considered. Therefore, knowing and monitoring these factors can help predict perceptual errors and reduce their probability. Furthermore, human condition monitoring (endogenous factor) is a task for passive brain–computer interfaces (BCIs) [5]. Unlike the traditional active BCI, which issues control commands through mental intent, passive BCIs continuously monitor the brain state during extended periods of cognitive activity and signal if it deviates from the normal state [6].

To control the quality of received information and its processing, the BCI must track both exogenous and endogenous components. Thus, the BCI must meet the task requirements (exogenous factor) along with the neural activity (endogenous factor). Moreover, objective assessments of the task requirements may depend on the amount of information, its ambiguity, and multimodality. Subjective estimates can be derived from the observer's reaction, such as response time, eye movements, and other behavioral indicators [7]. To take into account all these processes, it is necessary to move from a passive to a reactive BCI. The latter uses stimuli and analyzes the brain state through time intervals assigned to them [8]. Following this concept, a reactive BCI should analyze the brain state while performing a task. This will provide information on the influence of endogenous and exogenous factors on the speed and quality of information processing.

The further development of BCIs aims not only at the detection, but also the prediction of the human states. These BCIs will give rise to the artificial intelligence systems that assist or alarm when detecting a high probability of critical errors. Developing such systems requires finding the associations between the current state of the BCI operator and their performance in solving ongoing tasks. A bulk of literature associates changes in the human condition with their behavioral performance in ongoing tasks. In particular, attention, a fundamental aspect of the observer's state, modulates prestimulus alpha- and beta-band power [9–11], influencing the accuracy of perceptual decisions. Thus, either medium or low alpha- and high beta-band power during the prestimulus period is beneficial for sensory perception [11,12]. According to [13], the power and the prestimulus EEG phase coupling in the alpha- and beta-bands affect visual perception performance. Namely, better performance is associated with low phase coupling in the alpha-band and high phase coupling in the beta-band. Recent work [14] revealed that EEG power in the beta-2 frequency band at rest negatively correlated with the response times in the ongoing attentional task. While most works reported correlations between the neural correlates averaged across the trials, or between event-related potentials, in recent work [15], the authors used EEG power in different bands to predict individual performance in single trials contributing to the BCI problem.

Complementing the existing literature, we examined how the prestimulus EEG power predicts behavioral performance depending on the task demands. We considered a long-lasting monotonous experiment in which the participant perceived ambiguous stimuli and reported on each stimulus interpretation with the joystick buttons. The visual stimulus was an ambiguous Necker cube. The inner edges contrast defines one of two possible cube's orientations, left or right, and determines stimulus ambiguity. When ambiguity is low, cubes morphology is different for the left and right orientations. Therefore, subjects easily report the correct one. For the high ambiguity, stimulus morphology becomes similar for different orientations; therefore, subjects spent more effort to find the differences. In the recent works, we observed that subjects responded faster to the Necker cubes presented at the end of the experiment [16]. We also found that the brain utilized different neural mechanisms when processing stimuli with low and high ambiguity [2,4,17]. Based on these results, we hypothesized that during a long experiment with the Necker cubes, the observer's state changed, causing changes in behavioral performance. We expected to find the neural correlates of these changes in the prestimulus state and use them to predict the performance of the ongoing stimulus. We also supposed that changes in the human condition had a different effect depending on the stimulus ambiguity.

To test this hypothesis, we tracked the behavioral characteristics (decision times and errors) and simultaneously detected the EEG signals during a long monotonous task, including the Necker cubes interpretations. Behavioral monitoring revealed that decision time decreased with time on task, despite the ambiguity. For high ambiguity, we also observed a reduction of perceptual errors. EEG analysis showed growing prestimulus 9–11 Hz EEG power in the right temporal region. This EEG power negatively correlated with the decision time to the stimuli, with low ambiguity and the erroneous responses rate to stimuli with high ambiguity. The obtained results confirm that monitoring prestimulus EEG power enables predicting perceptual performance on the behavioral level.

2. Materials and Methods

2.1. Participants

Twenty healthy volunteers (nine females, 26–35 y.o.) with normal or corrected-to-normal vision participated in the experiments after providing written informed consent. Participants took part in similar experiments not earlier than six months before. All experiments were carried out in accordance with the requirements of the Declaration of Helsinki and approved by the local Research Ethics Committee of the Innopolis University.

2.2. Visual Stimuli

We used an experimental paradigm with an ambiguous bistable visual stimulus in the form of the Necker cube, which allows two possible interpretations [4,16,18,19]. The non-perceptually impaired volunteer interpreted this two-dimensional (2D) image as a three-dimensional (3D) object which is oriented either left or right. The balance between the brightness of three inner lines (1,2,3) located in the left bottom corner and three inner lines (4,5,6) in the right upper corner determines the ambiguity and orientation of the 3D cube (Figure 1A). The contrast parameter $a \in [0,1]$ is the normalized brightness of the inner lines (1,2,3) in the grayscale palette. In turn, the normalized brightness of the other inner lines (4,5,6) is defined as $1 - a$. Thus, the limiting cases $a = 0$ and $a = 1$ correspond to unambiguous 2D projections of the cube oriented to the left or to the right, respectively, whereas $a = 0.5$ implies a completely ambiguous spatial orientation of the 3D cube.

In the experiment, we used a set of the Necker cube images with $a = \{0.15, 0.25, 0.4, 0.45, 0.55, 0.6, 0.75, 0.85\}$ (Figure 1B), which we divided by subsets of cubes oriented to the left $a = \{0.15, 0.25, 0.4, 0.45\}$ and to the right $a = \{0.55, 0.6, 0.75, 0.85\}$. At this set, stimuli with low ambiguity (LA) $a = \{0.15, 0.25, 0.75, 0.85\}$, are easily interpreted, whereas the interpretation of stimuli with high ambiguity (HA) $a = \{0.40, 0.45, 0.55, 0.60\}$ requires more effort [4]. We also assume that HA processing engages more top-down control [20].

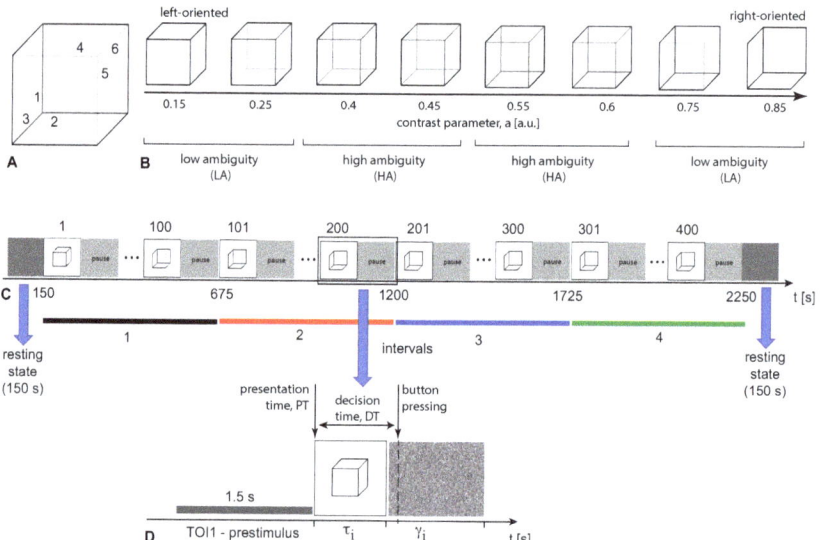

Figure 1. Experimental paradigm: (**A**) An example of the Necker cube image with the labeled inner edges. (**B**) Visual stimuli (Necker cubes) with different values of the contrast parameter a, which determines orientation and ambiguity. (**C**) Experimental protocol including 150-s resting state recordings and presentation of 400 stimuli alternating with the pauses. Colored horizontal stripes indicate 575-s time intervals. These intervals equally divide the stimuli presentation session. Each interval includes 100 stimuli. (**D**) Detailed illustration of a single stimulus presentation and abstract image. The cube presentation starts at the presentation time PT and lasts $\tau_i \in [1, 1.5]$ s. The decision time (DT) is determined by the interval between PT and button pressing. The pause time γ_i varies from 3 to 5 s. The time interval of interest (TOI1) is the 1.5-s pre-stimulus segment time-locked to the PT.

2.3. Experimental Protocol

Necker cubes (22.55 × 22.55 cm) were shown on a white background using a 24″ monitor (52.1 × 29.3 cm) with a 1920 × 1080 pixels resolution and a 60 Hz refresh rate. The distance between the participant and the monitor was 0.79—0.8 m, and the visual angle was ∼0.39 rad.

The duration of the entire experiment was about 40 min for each participant, and included EEG recordings of the eyes-open resting state (≈150 s) before and after the main part of the experiment. Cubes with predefined a values (selected from the set in Figure 1B) were randomly presented 400 times during the experimental session. Each cube with a particular ambiguity appeared about 50 times.

Each i-th stimulus presentation lasted for time interval τ_i, which ranged from $\tau_{min} = 1$ s to $\tau_{max} = 1.5$ s. The pauses between subsequent presentations of the Necker cube images, γ_i, ranged from $\gamma_{min} = 3$ s to $\gamma_{max} = 5$ s (Figure 1D) and contained an abstract image demo. The abstract image was the white noise picture (Figure 1D).

2.4. Behavioral Estimates

The participants were instructed to press either the left or right key in response to the left or right stimulus orientation, respectively. For each stimulus, we registered presentation time (PT)—the time between the beginning of the experiment and the moment when the current stimulus appeared on the screen. The behavioral response to each stimulus was assessed by measuring the decision time (DT), which corresponded to the time passed from the stimulus presentation to the button pressing (Figure 1C). We also monitored the correctness using error rate (ER) by comparing the actual stimulus orientation with the subject's response. The actual orientation of the Necker cube was defined by the contrast of the inner edges. Thus, $a = \{0.15, 0.25, 0.4, 0.45\}$ defined the left-oriented cubes, while $a = \{0.55, 0.6, 0.75, 0.85\}$ stood for the right-oriented ones. To define the correctness, we checked whether the subject pressed the left button for $a = \{0.15, 0.25, 0.4, 0.45\}$, or the right button for $a = \{0.55, 0.6, 0.75, 0.85\}$. Otherwise, their response was considered as incorrect. We excluded two subjects with $ER > 20\%$, as they exceeded the 90th percentile of ER distribution in the group.

2.5. EEG Recording

For registration of EEG signals, a monopolar method and a classical extended 10–10 electrode scheme were used. We recorded signals from 31 channels using an electrode cap, with two reference electrodes on the earlobes ($A1$ and $A2$) and a ground electrode N above the forehead. Ag/AgCl cup adhesive electrodes placed on the "Tien-20" paste (Weaver and Company, Aurora, CO, USA) were used for signal acquisition. Immediately before the experiments, a special abrasive "NuPrep" gel (Weaver and Company, Aurora CO, USA) was applied to the electrode attachment areas to increase skin conductivity. We maintained the impedance values in the range of 2–5 kΩ. For registration, amplification, and analog-to-digital conversion of the EEG signals, we used a multichannel electroencephalograph "Encephalan-EEG-19/26" (Medicom MTD company, Taganrog, Russian Federation) with a two-button input device (keypad). This device holds the registration certificate from the Federal Service for Supervision in Health Care No. FCP 2007/00124 of 07.11.2014 and European Certificate CE 538571 from the British Standards Institute (BSI).

The raw EEG signals were filtered by a fourth-order Butterworth (1–100)-Hz band-pass filter and a 50-Hz notch filter with built-in acquisition hardware and software. In addition, we performed an independent component analysis (ICA) to remove eye blinking and heartbeat artifacts. To determine components with artifacts, we examined their scalp map projections, waveforms, and spectra. The components containing eye-blinking artifacts usually had leading positions in the component array due to high amplitude. They demonstrated a smoothly decreasing spectrum, and their scalp map showed a strong far-frontal projection. Finally, eye-blinking artifacts had the typical waveform; therefore, those segments of EEG signals were marked by the experienced neurophysiologist, and used for determining the corresponding independent components.

We then segmented the EEG signals into 4-s trials, where each trial was associated with a single presentation of the Necker cube, and included a 2-s interval before and 2-s interval after the moment of the stimulus demonstration. After the EEG pre-processing procedure, we excluded some trials due to the remaining large-amplitude artifacts. To exclude trials containing large amplitude artifacts, we used the z-value threshold $z < 1$. The rejection procedure was performed using FieldTrip toolbox in Matlab [21].

After all preprocessing procedures, we had 52 ± 11 SD trials for the interval 1, 47 ± 11 SD trials for the interval 2, 47 ± 11 SD trials for the interval 3, and 55 ± 11 SD trials for the interval 4. We calculated the wavelet power for each trial in the (4–40)-Hz frequency range using the Morlet wavelet, and the number of cycles n was defined as $n = f$, where f is the signal frequency. Finally, we computed the event-related spectral perturbation (ERSP) by normalizing the wavelet power estimates W to the wavelet power of 40-s resting-state EEG as $ERSP = (W - W_{rest})/W_{rest}$. All processing procedures were performed offline.

Our goal was to study how the participant's state changed in the course of the experiment, regardless of the type of stimulus. Therefore, we measured brain activity before the start of the stimulus presentation (1.5-s prestimulus interval, TOI1 in Figure 1D).

2.6. Source Localization

We applied low-resolution precision electromagnetic brain tomography (eLORETA) to solve the inverse problem and localize the sources of neuronal activity according to EEG data at each of the predetermined points (voxels) in the brain volume [22–24].

LORETA is low-resolution brain electromagnetic tomography. This method solves the inverse problem: converting EEG measurements into information about the distribution of neural sources power into a brain volume. This method belongs to the class of nonparametric methods [25], which are based on the assumption that a separate current dipole (a source) is assigned to each of tens of thousands of elements of the tessellation of the cerebral cortex, while the orientation of the dipole is determined by the local normal to the surface. In this case, the inverse problem is linear, since the only unknowns are the amplitudes of the dipoles. Exact low-resolution brain electromagnetic tomography (eLORETA) is 3D, regularized, and minimum norm-weighted inverse solution with theoretically accurate zero error localization, even in the presence of structured biological or measurement noise [22,25]. The "Colin27" brain MRI averaged template [26] was used to develop a three-layer (brain, skull, and scalp) head model based on a boundary element method (BEM) [27,28]. The sources space inside the brain consisted of 11,865 voxels. The location of the EEG electrodes corresponded to the template head shape.

We analyzed the source characteristics in the predefined time-frequency domain of interest, selected on the basis of sensor-level analysis. To do this, we reassigned the EEG signals to the total average, subtracted the mean, and filtered with a fourth-order Butterworth $[f_L, f_H]$-Hz band-pass filter, where f_L and f_H define the frequency domain of interest. Then, we performed time-lock averaging across the TOI1 trials and computed the covariance matrix. The inverse solution yielded estimates of the source power in each voxel, averaged over the selected TOI window. Finally, we normalized the obtained estimates of the power P of each source to the power of 40-s EEG resting state EEG as $(P - P_{rest})/P_{rest}$. We used the automated anatomical labeling (AAL) brain atlas [29] to map the location of sources to the anatomical brain regions.

2.7. Experimental Conditions

Since the observer's state, as a rule, is not at a constant level, but fluctuates at different time scales, in the short term, this causes a difference in the subject's behavior when presenting stimuli, even if their ambiguity does not change. Meanwhile, in the long term, cognitive fatigue occurs, and the training effect takes place. In this work, we have eliminated short-term fluctuations and focused on the long-term changes in the person's condition over a 40-min task. To exclude the influence of short-scale fluctuations, we divided the entire experiment into four consecutive intervals of 10-min duration each (see Figure 1B). For each interval, we averaged ERSP and source power (SP) over all trials. To assess the behavioral characteristics, we used the median DT and the error rate (ER), reflecting the percentage of erroneous responses at each interval.

2.8. Statistical Testing

We tested how DT and ER changed at 1–4 intervals using additional controls of stimulus orientation and ambiguity. We performed repeated measures ANOVA with 1–4 intervals, ambiguity (HA and LA), and orientation (Left and Right) as within-subject factors. In general, ANOVA requires the homogeneity assumption: the population variances of the dependent variable must be equal for all groups. At the same time, this assumption may be ignored if the sample size is equal for each group. In this study, we used a within-subject design with repeated measures. Therefore, the sample size for each condition was equal, and we did not control for variance homogeneity. If the tested samples did not obey the normality condition, we applied Greenhouse–Geisser correction to ANOVA results. For significant main effects, we performed a post hoc analysis using parametric or nonparametric tests, depending on sample normality, which was determined using the Shapiro–Wilk test. All test types are specified in the Result section and in the figures captions. A statistical analysis was performed in IBM SPSS Statistics.

Statistical analyses of brain activity were carried out based on the subject-level wavelet power, averaged over trials and over TOI1. Contrasts between the four intervals were tested for statistical significance using a permutation test combined with the cluster-based correction for multiple comparisons. Specifically, the F-tests compared four wavelet power sets for all pairs (channel, frequency). Items that passed the threshold corresponding to a p-value of 0.001 (one-tailed) were labeled along with their adjacent items and collected in separate negative and positive clusters. The minimum required number of neighbors was set to 2. The F-values in each cluster were summarized and corrected. The maximum amount was entered into the permutation structure as a test statistic. A cluster was considered significant if its p-value was below 0.01. The number of permutations was 2000.

A similar procedure was followed on the source level results. We performed a cluster-corrected statistical intra-subject permutation test on the test-averaged and TOI1-averaged source power distributions to determine significant differences between four intervals [30,31]. The threshold for paired comparisons with F-test was $p = 0.005$. The p-threshold for the cluster was 0.025. The number of permutations was 2000. Finally, we calculated, for each subject, the average power of the source activity in the region of the identified cluster for each of the four intervals.

All described operations were performed in Matlab using the Fieldtrip toolbox [21,32].

3. Results

3.1. Results of the Behavioral Data Analysis

Contrasting subjects' DTs on four intervals, we observed significant changes among intervals, the effect of stimulus ambiguity, and the combined effect of ambiguity and orientation (see Table 1). Nevertheless, the experiment demonstrated that DT varies over the course of the experiment in the same way for all stimuli. We also found a significant effect of ambiguity on ER. At the same time, we observed a significant interaction effect of interval and ambiguity (see Table 2). These results suggest that ER differs for HA and LA stimuli regardless of their orientation. Moreover, ER changed differently during the experiment depending on the ambiguity.

Table 1. Median decision time to the current stimulus, DT [s] (ANOVA Summary).

Factors	dF_1	dF_2	F	p
Interval	2.285	38.853	5.805	0.005 *
Ambiguity	1	17	79.524	<0.0001 *
Orientation	1	17	1.093	0.310
Interval × Ambiguity	3	51	1.206	0.317
Interval × Orientation	3	51	1.290	0.288
Ambiguity × Orientation	1	17	6.385	0.022 *
Interval × Ambiguity × Orientation	3	51	0.544	0.655

Here, '*' indicates the level of significance $p < 0.05$.

Table 2. Percentage of erroneous responses to the current stimulus, ER [%] (ANOVA Summary).

Factors	dF_1	dF_2	F	p
Interval	1.995	33.910	2.988	0.064
Ambiguity	1	17	13.128	0.002 *
Orientation	1	17	2.671	0.121
Interval × Ambiguity	3	51	5.918	0.002 *
Interval × Orientation	1.975	33.58	1.454	0.248
Ambiguity × Orientation	1	17	0.922	0.350
Interval × Ambiguity × Orientation	1.63	27.715	1.892	0.175

Here, '*' indicates the level of significance $p < 0.05$.

The post hoc analysis revealed that subjects responded faster to LA stimuli than HA ones: $Z = -3.724, p < 0.0001$, Wilcoxon test (Figure 2A). For HA stimuli, DT was similar for the left- and right-oriented stimuli: $t(17) = 0.383, p = 0.706$, t-test (Figure 2B). For LA stimuli, subjects responded faster to the left-oriented stimuli: $Z = -2.591, p = 0.01$, Wilcoxon test (Figure 2C). Analysis of pairwise differences displayed that 12 subjects demonstrated effects in the same direction as the group. Finally, DT decreased with the interval: $\chi^2(3) = 12.218, p = 0.007$, Friedman test (Figure 2D). The post hoc Wilcoxon test revealed higher ER for HA stimuli when compared to LA stimuli: $Z = -3.29, p = 0.001$ (Figure 2E). For HA stimuli, ER decreased during the course of the experiment: $\chi^2(3) = 7.545$, $p = 0.056$, Friedman test (Figure 2F). Finally, there was no correlation between age and DT to HA stimuli: $r(20) = -0.24, p = 0.3$ and LA stimuli: $r(20) = -0.31, p = 0.17$. DT was similar for males and females for both HA stimuli: $t(18) = 0.79, p = 0.436$ and LA stimuli: $t(18) = 0.96, p = 0.348$.

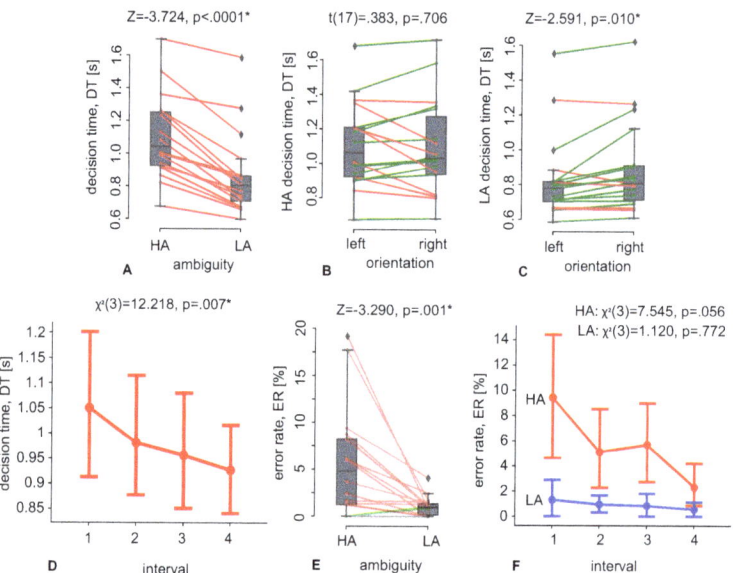

Figure 2. Results of the behavioral data analysis: (**A**) Median decision time, DT to HA and LA stimuli (*$p < 0.0001$ via Wilcoxon test, uncorrected). (**B**) Median DT to the left- and right-oriented HA stimuli ($p = 0.706$ via t-test, uncorrected). (**C**) Median DT to the left- and right-oriented LA stimuli (*$p = 0.01$ via Wilcoxon test, uncorrected). (**D**) Median DT (group mean ± 95% CI) on four intervals (*$p = 0.007$ via Friedman test, uncorrected). (**E**) Percentage of erroneous responses (ER) to the HA and LA stimuli (*$p = 0.001$ via Wilcoxon test, uncorrected). (**F**) ER (group means ± 95% CI) on all intervals separately for HA ($p = 0.056$ via Friedman test, uncorrected) and LA ($p = 0.772$ via Friedman test, uncorrected) stimuli.

3.2. Results of the EEG Data Analysis on the Sensor Level

Comparing the prestimulus ERSP on four intervals, we found one cluster with $p = 0.0015$ (corrected using the permutation statistics) in the 9–11 Hz frequency band, including parietal and temporal sensors (P8, TP8, T8, FC4, FT8) in the right hemisphere (Figure 3A). The ERSP averaged over the EEG sensors in this cluster grew with the interval number from 0.18 ± 0.11 SE at interval 1 to 0.49 ± 0.15 SE at interval 4 (Figure 3B). We also compared the ERSP between males and females and between two age groups ("<26 y.o." vs. ">26 y.o.", where 26 was the median age) via the independent-samples t-test with cluster-based correction for multiple comparisons. In both cases, the difference was insignificant.

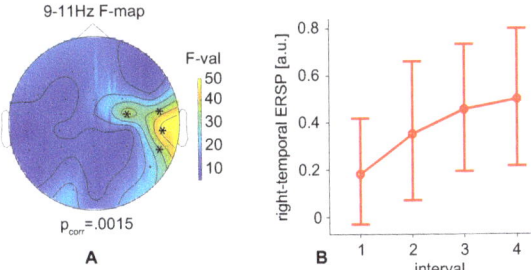

Figure 3. Results of the EEG data analysis on the sensor level: A scalp topogram illustrates F-value and EEG channels cluster, demonstrating the significant change of ERSP between four intervals (*$p = 0.0015$ via F-test and cluster-based correction for multiple comparisons) (**A**). Changing ERSP in this cluster (group mean ± 95% CI) with the time with the interval number (**B**).

3.3. Results of the EEG Data Analysis in the Source Space

Setting the bandwidth of interest to 10 ± 2 Hz, we contrasted the source power (SP) between the four intervals. The statistical analysis revealed one cluster in the source space with $p = 0.02$ (corrected using the permutation statistics). This cluster included voxels in the right middle temporal gyrus (Temporal Mid R), right superior temporal gyrus (Temporal Sup R), right inferior temporal gyrus (Temporal Inf R), Rolandic operculum (Rolandic Oper R), fusiform gyrus (Fusiform R), and a part of the cerebellum (Cerebellum Crus1 R) (Figure 4A). The maximal F-value was achieved in the right middle temporal gyrus, while the minimum F-value belonged to the cerebellum. The averaged SP in this cluster grew from interval 1 (-0.22 ± 0.12 SD) to interval 4 (1.21 ± 0.49 SD) (Figure 4B).

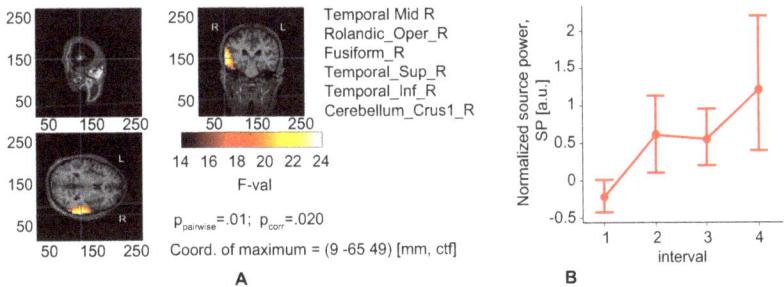

Figure 4. Results of the EEG data analysis in the source space: (**A**) Source plot shows F-value, reflecting the significant change of the source power (SP) between four intervals on the prestimulus interval $t \in [-1.5, 0]$ s, for $f \in 8 - 12$ Hz ($p = 0.02$ via F-test, permutation-based correction). Legends contain p-values, CTF coordinates of the voxel with maximal F-value, and names of anatomical zones according to Automated Anatomical Labeling (AAL); (**B**) Normalized source power, NSP (group mean ± 95% CI) in this cluster on four intervals.

3.4. Results of the Correlation Analysis

Using repeated measures correlation analysis, we found that SP was negatively correlated with DT to LA stimuli (Figure 5A) and ER for HA stimuli (Figure 5D). At the same time, there was no correlation between SP and DT to HA stimuli (Figure 5B), or between SP and ER for LA stimuli (Figure 5C).

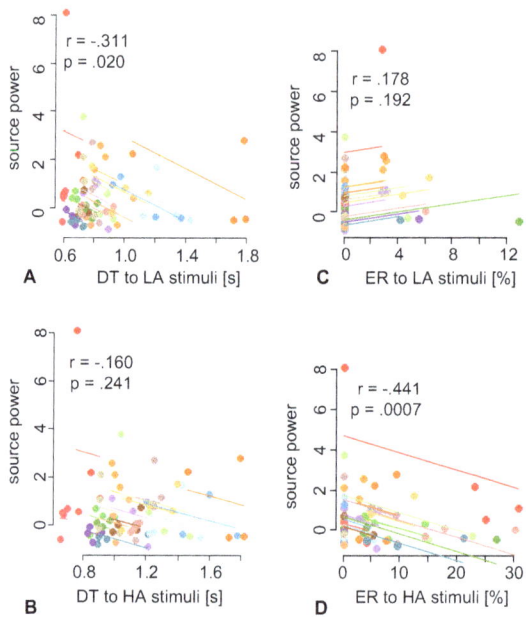

Figure 5. Results of the correlation analysis: Regression plots illustrate the relationship between SP and DT to LA stimuli (**A**); SP and DT to HA stimuli (**B**), SP and ER for LA stimuli (**C**); SP and ER for Ha stimuli (**D**). The colored dots correspond to each participant's data; the lines have the same slope estimated for these participants via correlation analysis with repeated measures.

4. Discussion

A group of volunteers was tasked to classify Necker cubes of different ambiguity within 40 min. The subjects reported the orientation (left or right) of each presented cube, while the stimulus morphology ranged from low ambiguity (LA) to high ambiguity (HA). By observing behavioral responses, we found that decision time decreased with the time on task for both HA and LA stimuli (Figure 2D). At the same time, the subjects improved the correctness of the interpretation of the HA stimuli, but not the LA stimuli (Figure 2F). Analysis of the EEG spectral power on the sensor level and in the source space revealed an increase in the prestimulus power at 9–11 Hz with the time on the task (Figures 3B and 4B). Finally, we found that the prestimulus EEG power negatively correlated with the decision time to LA stimuli (Figure 5A) and the number of erroneous responses to HA (Figure 5D) stimuli.

First, we hypothesized that the prestimulus EEG power reflects changes in a person's condition. The condition, in turn, affected the performance of processing the ongoing visual stimulus [11,16]. Thus, our results showed that the high pre-stimulus 9–11 Hz EEG power predicted faster decision times and greater accuracy. It is worth noting that we compared EEG power and behavioral estimates between time segments, each of which lasted 10 min. Therefore, we have associated the described effects with slow changes in the observer's state. Taken together, we proposed a possible application of our findings in passive brain-computer interfaces to monitor the human's condition and predict decision speed and errors.

Along with possible practical applications, our result can help to reveal specific features of ambiguous perception. To start a discussion on this aspect, we will look at the limitations of our experimental design. It consists of two confounding variables, mental fatigue, and learning; both can be time-dependent. Thus, the revealed EEG changes practically do not reflect one of these factors. At the same time, we suppose that fatigue and learning affect the observer state regardless of the stimulus ambiguity. In contrast, our results show a different time-on-task effect on processing HA and LA stimuli. For instance, we revealed a reduction in error rate for HA stimuli only. In addition, the prestimulus EEG power negatively correlates with the decision time to LA stimuli, not to HA stimuli, and negatively correlates with the error rate for HA, but not for LA stimuli. Another limitation is the small sample size. Thus, there is a risk that the individual characteristics of the participants, e.g., gender and age, will affect their perception of ambiguous stimuli. For our group, we observed no gender and age effects on the decision time and the ERSP, due to the almost uniform distribution of these factors between participants. At the same time, we expect that another group of younger or older subjects may demonstrate different effects on the behavioral and neural activity levels.

The decision time (DT) may decrease due to neural adaptation (NA), which occurs when the same visual stimulus is repeatedly presented within a short interval and causes a decrease in neural response to repetitive versus non-repeated stimulus [33]. The NA is thought to arise from at least two types of neural activity. One explanation is that only the part belonging to the neuronal ensemble is sensitive to stimulus recognition. Thus, the neurons that are not critical for stimulus recognition decrease their responses when the stimulus reappears, while on the contrary, neuronal populations carrying essential information continue to give a robust response. As a result, the mean firing rate decreases due to stimulus repetition [34]. An alternative explanation is that stimulus repetition reduces response in the time domain [33]. According to this theory, a neural network that processes sensory information responds faster to a repetitive stimulus than to a new stimulus, i.e., a stable response. The network connections involved in the response creation were reinforced by the previous presentation of the same stimulus [35]. The NA affects the neuronal response in the occipital [36], parietal [37], and frontal [38] cortex areas in single-unit data, and on the sensory level. As known from the literature, NA, as a rule, reduces the stimulus-related EEG/ECoG response to stimuli. Here, we did not consider the post-stimulus EEG power and did not report such signs of NA. At the same time, an increase in the prestimulus EEG power may reflect the preactivation of sensory neurons. We suppose that in this preparatory state the neural ensemble exhibits less activation in response to the stimulus. Further research should verify this hypothesis by examining the post-stimulus activity as a function of the time-on-task.

Another potential explanation is the vital role of alpha-band oscillations for visual perception. Our results showed that an increase in the 9–11-Hz band power correlates with enhancing processing performance. In contrast, many studies have reported negative effects of alpha power on processing performance. For example, the authors of Ref. [12] reported that moderate to low alpha signal strength in the prestimulus period is beneficial for sensory perception. In contrast, a recent review [39] emphasizes that high alpha power facilitates perception by suppressing irrelevant input and generating predictors in the visual cortex. The latter was confirmed by observing an increase in the prestimulus alpha power when participants could predict the identity of the forthcoming stimulus [40].

Finally, the role of alpha-band oscillations depends largely on their incident brain region. For instance, the authors of Ref. [41] provided evidence that right temporal alpha oscillations play a crucial role in inhibiting habitual thinking modes, thereby developing creative cognition. Another work showed that observing a Necker cube can improve subsequent creative problem-solving [42]. In line with these works, we supposed that increasing 9–11 Hz power in the right temporal region reflects a developing ability to inhibit obvious associations. According to Ref. [42], the latter may be a biomarker of neural processes facilitating creative problem-solving.

We hypothesize that NA affects the bottom-up processing, while the other two represent the top-down processing components [43,44]. Assuming that NA facilitates the bottom-up processing, we provide a possible explanation for the negative correlation between the prestimulus EEG power and the decision time to LA stimuli. The morphology of inner edges unambiguously determines the orientation of the LA stimulus. Consequently, during the LA stimulus processing, the bottom-up component prevails [45]. If so, neurons in sensory areas receive the information needed to make the right decision. Repeated stimulation pre-activates these sensory neurons, which leads to a decrease in decision time to LA stimuli. In contrast, the HA stimuli processing may require top-down mechanisms, because the morphology remains similar for the cube being either left or right-oriented [20]. To explain the lack of correlation between the prestimulus EEG power and decision time to HA stimuli, we also assume that during HA processing, the top-down component predominates for most of the time interval, while the bottom-up one may be limited to an earlier and shorter time window [16]. Thus, by facilitating the bottom-up processing, NA has little or no effect on the overall decision time for HA stimuli.

Summing up, we can say that the increasing prestimulus EEG power can reflect NA in sensory neural networks encoding the Necker cube morphology, which explains the negative correlation between the EEG power and the decision time to LA stimuli. At the same time, decision time to HA stimuli also decreases with time on task, but barely correlates with the prestimulus EEG power. This behavioral effect is probably the result of neural processes acting after the stimulus onset and relies on the integrative dynamics, rather than the EEG power modulation in a particular area. In our future studies, we will address this issue by considering the functional connectivity evolution during the experiment.

By examining the error rate, we found that the observer responded more correctly to LA stimuli. The number of erroneous responses is less than 2%, and remains unchanged during the experiment. In contrast, the number of incorrect responses to HA stimuli decreased with time on task and negatively correlated with the prestimulus 9–11 Hz EEG power. This effect may be a result of the top-down mechanisms, e.g., predicting the identity of the forthcoming stimulus or creative thinking. As discussed above, these processes also accompany increasing alpha-band power and relate to ambiguous stimuli processing.

5. Conclusions

During the Necker cube classification, participants decrease their decision time with the time on task, regardless of the stimulus ambiguity. At the same time, they improve the correctness of interpretation only for the highly ambiguous stimuli. EEG analysis has revealed growing prestimulus alpha-band power with the time on task. We have found that the prestimulus EEG power negatively correlates with the decision time for the stimuli, with low ambiguity and the number of erroneous responses to highly ambiguous stimuli.

We suppose that repetitive stimuli presentation affects top-down and bottom-up processing mechanisms. Thus, it may cause the neuronal adaptation of the sensory neurons facilitating bottom-up processing. For the low ambiguity, the bottom-up component dominates; therefore, decision time correlates with the prestimulus EEG power. Increasing alpha-band EEG power may also reflect modulation of top-down components, e.g., the ability to suppress irrelevant information and form the predictors of ongoing stimulus. For the high ambiguity, the top-down processes dominate; therefore, the correctness rate correlates with the prestimulus EEG power.

Finally, our results may add to uncovering the associations between the current human condition and their performance in solving ongoing tasks. It is essential for BCIs to aim not only at the detection, but also the prediction of the human states. These BCIs will give rise to the artificial intelligence systems that assist or alarm when detecting a high probability of critical errors.

Author Contributions: Conceptualization, A.N.P.; Data curation, A.E.H.; Formal analysis, S.A.K. and A.N.P.; Funding acquisition, A.E.H.; Investigation, A.K.K. and S.A.K.; Methodology, V.A.M. and A.E.H.; Project administration, A.E.H.; Software, A.K.K.; Validation, V.A.M.; Visualization, S.A.K. and V.A.M.; Writing—original draft, V.A.M.; Writing—review and editing, A.N.P. and A.E.H. All authors have read and agreed to the published version of the manuscript.

Funding: This research was funded by the Russian Science Foundation (Project 19-72-10121) V.A.M. supported by the Russian Foundation for Basic Research (19-32-60042) in part of the experimental paradigm development and behavioral data analysis. S.A.K. is supported by the President Program (MD-1921.2020.9) in the part of source reconstruction.

Institutional Review Board Statement: The study was conducted according to the guidelines of the Declaration of Helsinki, and approved by the Institutional research ethics committee of the Innopolis University (protocol 2 from 18 January 2019).

Informed Consent Statement: Informed consent was obtained from all subjects involved in the study.

Acknowledgments: The authors graciously acknowledge the subjects for their contribution to this research.

Conflicts of Interest: The authors declare no conflict of interest. The funders had no role in the design of the study; in the collection, analyses, or interpretation of data; in the writing of the manuscript, or in the decision to publish the results.

References

1. Heekeren, H.R.; Marrett, S.; Bandettini, P.A.; Ungerleider, L.G. A general mechanism for perceptual decision-making in the human brain. *Nature* **2004**, *431*, 859–862. [CrossRef]
2. Maksimenko, V.; Kuc, A.; Frolov, N.; Kurkin, S.; Hramov, A. Effect of repetition on the behavioral and neuronal responses to ambiguous Necker cube images. *Sci. Rep.* **2021**, *11*, 3454. [CrossRef]
3. Summerfield, C.; De Lange, F.P. Expectation in perceptual decision making: Neural and computational mechanisms. *Nat. Rev. Neurosci.* **2014**, *15*, 745–756. [CrossRef]
4. Maksimenko, V.A.; Kuc, A.; Frolov, N.S.; Khramova, M.V.; Pisarchik, A.N.; Hramov, A.E. Dissociating Cognitive Processes During Ambiguous Information Processing in Perceptual Decision-Making. *Front. Behav. Neurosci.* **2020**, *14*, 95. [CrossRef]
5. Aricò, P.; Borghini, G.; Di Flumeri, G.; Sciaraffa, N.; Babiloni, F. Passive BCI beyond the lab: Current trends and future directions. *Physiol. Meas.* **2018**, *39*, 08TR02. [CrossRef]
6. Arico, P.; Borghini, G.; Di Flumeri, G.; Sciaraffa, N.; Colosimo, A.; Babiloni, F. Passive BCI in operational environments: Insights, recent advances, and future trends. *IEEE Trans. Biomed. Eng.* **2017**, *64*, 1431–1436. [CrossRef] [PubMed]
7. Hramov, A.E.; Maksimenko, V.A.; Pisarchik, A.N. Physical principles of brain-computer interfaces and their applications for rehabilitation, robotics and control of human brain states. *Phys. Rep.* **2021**, *918*, 1–133. [CrossRef]
8. Zander, T.O.; Kothe, C. Towards passive brain–computer interfaces: Applying brain–computer interface technology to human–machine systems in general. *J. Neural Eng.* **2011**, *8*, 025005. [CrossRef] [PubMed]
9. Anderson, K.; Ding, M. Attentional modulation of the somatosensory mu rhythm. *Neuroscience* **2011**, *180*, 165–180. [CrossRef] [PubMed]
10. Bauer, M.; Kennett, S.; Driver, J. Attentional selection of location and modality in vision and touch modulates low-frequency activity in associated sensory cortices. *J. Neurophysiol.* **2012**, *107*, 2342–2351. [CrossRef] [PubMed]
11. Gola, M.; Magnuski, M.; Szumska, I.; Wróbel, A. EEG beta band activity is related to attention and attentional deficits in the visual performance of elderly subjects. *Int. J. Psychophysiol.* **2013**, *89*, 334–341. [CrossRef]
12. Van Dijk, H.; Schoffelen, J.M.; Oostenveld, R.; Jensen, O. Prestimulus oscillatory activity in the alpha band predicts visual discrimination ability. *J. Neurosci.* **2008**, *28*, 1816–1823. [CrossRef]
13. Hanslmayr, S.; Aslan, A.; Staudigl, T.; Klimesch, W.; Herrmann, C.S.; Bäuml, K.H. Prestimulus oscillations predict visual perception performance between and within subjects. *Neuroimage* **2007**, *37*, 1465–1473. [CrossRef]
14. Rogala, J.; Kublik, E.; Krauz, R.; Wróbel, A. Resting-state EEG activity predicts frontoparietal network reconfiguration and improved attentional performance. *Sci. Rep.* **2020**, *10*, 5064. [CrossRef]
15. Vecchio, F.; Alù, F.; Orticoni, A.; Miraglia, F.; Judica, E.; Cotelli, M.; Rossini, P.M. Performance prediction in a visuo-motor task: The contribution of EEG analysis. *Cogn. Neurodyn.* **2021**, 1–12. [CrossRef]
16. Maksimenko, V.A.; Frolov, N.S.; Hramov, A.E.; Runnova, A.E.; Grubov, V.V.; Kurths, J.; Pisarchik, A.N. Neural Interactions in a Spatially-Distributed Cortical Network During Perceptual Decision-Making. *Front. Behav. Neurosci.* **2019**, *13*, 220. [CrossRef] [PubMed]
17. Kuc, A.; Grubov, V.V.; Maksimenko, V.A.; Shusharina, N.; Pisarchik, A.N.; Hramov, A.E. Sensor-Level Wavelet Analysis Reveals EEG Biomarkers of Perceptual Decision-Making. *Sensors* **2021**, *21*, 2461. [CrossRef] [PubMed]
18. Wang, M.; Arteaga, D.; He, B.J. Brain mechanisms for simple perception and bistable perception. *Proc. Natl. Acad. Sci. USA* **2013**, *110*, E3350–E3359. [CrossRef] [PubMed]
19. Kornmeier, J.; Friedel, E.; Wittmann, M.; Atmanspacher, H. EEG correlates of cognitive time scales in the Necker-Zeno model for bistable perception. *Conscious. Cogn.* **2017**, *53*, 136–150. [CrossRef]
20. Engel, A.K.; Fries, P. Beta-band oscillations—Signalling the status quo? *Curr. Opin. Neurobiol.* **2010**, *20*, 156–165. [CrossRef]

21. Oostenveld, R.; Fries, P.; Maris, E.; Schoffelen, J.M. FieldTrip: Open source software for advanced analysis of MEG, EEG, and invasive electrophysiological data. *Comput. Intell. Neurosci.* **2011**, *2011*, 156869. [CrossRef]
22. Pascual-Marqui, R.D. Discrete, 3D distributed, linear imaging methods of electric neuronal activity. Part 1: Exact, zero error localization. *arXiv* **2007**, arXiv:0710.3341.
23. Pascual-Marqui, R.D.; Biscay-Lirio, R.J. Interaction patterns of brain activity across space, time and frequency. Part I: Methods. *arXiv* **2011**, arXiv:1103.2852.
24. Pascual-Marqui, R.D. Theory of the EEG inverse problem. In *Quantitative EEG Analysis: Methods and Clinical Applications*; Artech House: Norwood, MA, USA, 2009; pp. 121–140.
25. Grech, R.; Cassar, T.; Muscat, J.; Camilleri, K.P.; Fabri, S.G.; Zervakis, M.; Xanthopoulos, P.; Sakkalis, V.; Vanrumste, B. Review on solving the inverse problem in EEG source analysis. *J. Neuroeng. Rehabil.* **2008**, *5*, 25. [CrossRef]
26. Holmes, C.J.; Hoge, R.; Collins, L.; Woods, R.; Toga, A.W.; Evans, A.C. Enhancement of MR images using registration for signal averaging. *J. Comput. Assist. Tomogr.* **1998**, *22*, 324–333. [CrossRef]
27. Fuchs, M.; Kastner, J.; Wagner, M.; Hawes, S.; Ebersole, J.S. A standardized boundary element method volume conductor model. *Clin. Neurophysiol.* **2002**, *113*, 702–712. [CrossRef]
28. Baillet, S.; Mosher, J.C.; Leahy, R.M. Electromagnetic brain mapping. *IEEE Signal Process. Mag.* **2001**, *18*, 14–30. [CrossRef]
29. Tzourio-Mazoyer, N.; Landeau, B.; Papathanassiou, D.; Crivello, F.; Etard, O.; Delcroix, N.; Mazoyer, B.; Joliot, M. Automated anatomical labeling of activations in SPM using a macroscopic anatomical parcellation of the MNI MRI single-subject brain. *Neuroimage* **2002**, *15*, 273–289. [CrossRef]
30. Maris, E. Statistical testing in electrophysiological studies. *Psychophysiology* **2012**, *49*, 549–565. [CrossRef]
31. Maris, E.; Oostenveld, R. Nonparametric statistical testing of EEG-and MEG-data. *J. Neurosci. Methods* **2007**, *164*, 177–190. [CrossRef] [PubMed]
32. Chholak, P.; Kurkin, S.A.; Hramov, A.E.; Pisarchik, A.N. Event-Related Coherence in Visual Cortex and Brain Noise: An MEG Study. *Appl. Sci.* **2021**, *11*, 375. [CrossRef]
33. Henson, R.; Rugg, M. Neural response suppression, haemodynamic repetition effects, and behavioural priming. *Neuropsychologia* **2003**, *41*, 263–270. [CrossRef]
34. Wiggs, C.L.; Martin, A. Properties and mechanisms of perceptual priming. *Curr. Opin. Neurobiol.* **1998**, *8*, 227–233. [CrossRef]
35. Henson, R.N.; Price, C.J.; Rugg, M.D.; Turner, R.; Friston, K.J. Detecting latency differences in event-related BOLD responses: Application to words versus nonwords and initial versus repeated face presentations. *Neuroimage* **2002**, *15*, 83–97. [CrossRef] [PubMed]
36. Kourtzi, Z.; Kanwisher, N. Cortical regions involved in perceiving object shape. *J. Neurosci.* **2000**, *20*, 3310–3318. [CrossRef]
37. Naccache, L.; Dehaene, S. The priming method: Imaging unconscious repetition priming reveals an abstract representation of number in the parietal lobes. *Cereb. Cortex* **2001**, *11*, 966–974. [CrossRef] [PubMed]
38. Wagner, A.D.; Koutstaal, W.; Maril, A.; Schacter, D.L.; Buckner, R.L. Task-specific repetition priming in left inferior prefrontal cortex. *Cereb. Cortex* **2000**, *10*, 1176–1184. [CrossRef]
39. Clayton, M.S.; Yeung, N.; Cohen Kadosh, R. The many characters of visual alpha oscillations. *Eur. J. Neurosci.* **2018**, *48*, 2498–2508. [CrossRef]
40. Mayer, A.; Schwiedrzik, C.M.; Wibral, M.; Singer, W.; Melloni, L. Expecting to see a letter: Alpha oscillations as carriers of top-down sensory predictions. *Cereb. Cortex* **2015**, *26*, 3146–3160. [CrossRef] [PubMed]
41. Luft, C.D.B.; Zioga, I.; Thompson, N.M.; Banissy, M.J.; Bhattacharya, J. Right temporal alpha oscillations as a neural mechanism for inhibiting obvious associations. *Proc. Natl. Acad. Sci. USA* **2018**, *115*, E12144–E12152. [CrossRef] [PubMed]
42. Laukkonen, R.E.; Tangen, J.M. Can observing a Necker cube make you more insightful? *Conscious. Cogn.* **2017**, *48*, 198–211. [CrossRef] [PubMed]
43. Ding, N.; Simon, J.Z. Emergence of neural encoding of auditory objects while listening to competing speakers. *Proc. Natl. Acad. Sci. USA* **2012**, *109*, 11854–11859. [CrossRef] [PubMed]
44. Baldeweg, T. Repetition effects to sounds: Evidence for predictive coding in the auditory system. *Trends Cogn. Sci.* **2006**, *10*, 93–94.
45. Okazaki, M.; Kaneko, Y.; Yumoto, M.; Arima, K. Perceptual change in response to a bistable picture increases neuromagnetic beta-band activities. *Neurosci. Res.* **2008**, *61*, 319–328. [CrossRef] [PubMed]

Article

Classification of Event-Related Potentials with Regularized Spatiotemporal LCMV Beamforming

Arne Van Den Kerchove [1,2,*], Arno Libert [1], Benjamin Wittevrongel [1] and Marc M. Van Hulle [1]

[1] Laboratory for Neuro- and Psychophysiology, Department of Neuroscience, KU Leuven, Campus Gasthuisberg, O&N 2, Herestraat 49 Bus 1021, 3000 Leuven, Belgium; arno.libert@kuleuven.be (A.L.); benjamin.wittevrongel@kuleuven.be (B.W.); marc.vanhulle@kuleuven.be (M.M.V.H.)

[2] Centre de Recherche en Informatique, Signal et Automatique de Lille (CRIStAL), UMR 9189, Université de Lille, Campus Scientifique, Bâtiment ESPRIT, Avenue Henri Poincaré, 59300 Villeneuve-d'Ascq, France

* Correspondence: arne.vandenkerchove@kuleuven.be

Featured Application: Brain–computer interfaces as assistive technology to restore communication capabilities for disabled patients.

Abstract: The usability of EEG-based visual brain–computer interfaces (BCIs) based on event-related potentials (ERPs) benefits from reducing the calibration time before BCI operation. Linear decoding models, such as the spatiotemporal beamformer model, yield state-of-the-art accuracy. Although the training time of this model is generally low, it can require a substantial amount of training data to reach functional performance. Hence, BCI calibration sessions should be sufficiently long to provide enough training data. This work introduces two regularized estimators for the beamformer weights. The first estimator uses cross-validated L2-regularization. The second estimator exploits prior information about the structure of the EEG by assuming Kronecker–Toeplitz-structured covariance. The performances of these estimators are validated and compared with the original spatiotemporal beamformer and a Riemannian-geometry-based decoder using a BCI dataset with P300-paradigm recordings for 21 subjects. Our results show that the introduced estimators are well-conditioned in the presence of limited training data and improve ERP classification accuracy for unseen data. Additionally, we show that structured regularization results in lower training times and memory usage, and a more interpretable classification model.

Keywords: brain–computer interface; event-related potential; beamforming; regularization

1. Introduction

Brain–computer interfaces (BCIs) establish a direct communication pathway between the brain and an external device [1]. Severely disabled patients with impaired or absent communication capabilities can benefit from BCIs to restore normal functioning [2,3]. BCIs can be implemented in multiple ways, using non-invasive recording techniques such as electroencephalography (EEG) [4], magnetoencephalography (MEG) [5], functional near-infrared spectroscopy (fNIRS) [6], and optically pumped magnetometers (OPM MEG) [7], or semi-invasive and invasive methods such as electrocorticography (ECoG) [8] or microelectrode arrays [9], which require surgery to implant a recording device. Although invasive BCIs yield the highest information transfer rate [10], non-invasive BCIs are preferable for short-term use since they are not susceptible to the risks that come with surgery. Among the non-invasive options, EEG is the most cost-effective and practical as it is not limited to the same controlled settings as MEG and OPM MEG.

In addition to the recording method, BCIs differ in the communication paradigms used for communication [4]. A popular class of BCI paradigms relies on the evocation of event-related potentials (ERPs) in the brain in response to visual, auditory, or tactile stimulation, given their low decoding cost and generally short calibration time before usage [11,12]. In this study we focused on the visual P300 oddball ERP in response to a rare but attended visual stimulus. The decoder detects whether this ERP is present to determine which stimulus the user attended. The P300 paradigm has been used extensively in BCI development and is easy to set up [13–16].

There are multiple state-of-the-art P300 classification methods, such as support vector machines (SVMs) [17], deep learning models [18,19], and Riemannian geometry classifiers [15]. Although these models often return a high classification accuracy, there is a need for lightweight models, as lightweight models lead to fast off-line analyses and can be transferred to consumer-grade hardware. When moving towards plug-and-play solutions, BCI calibration sessions should be short and model training times should be low. The spatiotemporal beamformer [20,21] belongs to this class of ERP-decoding models as it achieves state-of-the-art performance and is fast to train. Earlier work has shown that it is possible to apply the spatiotemporal beamformer to multiple time-locked visual BCI paradigms, including the P300 oddball paradigm, steady-state visually evoked potentials (SSVEPs), code-modulated visually evoked potentials (cVEPs) [22], and motion-onset visually evoked potentials (mVEPs) [23].

This work shows that the original spatiotemporal beamformer [21] can fall short in its performance when BCI calibration data are restricted. We also show that the spatiotemporal beamformer does not scale well for higher spatial and temporal resolution cases. As a response to these issues, we introduce a regularization method that exploits prior knowledge about the spatiotemporal nature of the EEG signal to improve the accuracy for settings with low data availability and to speed up the classifier training time, thereby considerably reducing memory usage. Similar structured regularization approaches have been applied to other linear ERP classifiers [24,25] and have shown significant increases in performance. Additionally, we show that regularization results in an interpretable classification model, which can aid in analyzing and developing spatiotemporal beamformer-based classifiers.

2. Materials and Methods

2.1. Notation

We represent matrices with cursive capital letters, vectors with bold lowercase letters, and scalars with cursive lowercase letters. The epoched EEG data with n epochs, c channels, and s samples are represented in epoch format as $\{X_i \in \mathbb{R}^{c \times s}\}_{i=1}^{n}$ or in flattened vector format by concatenating all channels for each epoch. Flattening results in $\{\mathbf{x}_i \in \mathbb{R}^{cs}\}_{i=1}^{n}$ such that $\mathbf{x}_i = \text{vec}(X_i)$. The real covariance matrix of the epochs in vector format is denoted by C, and estimators thereof as $\hat{C}$.

2.2. Spatiotemporal Beamforming

LCMV-beamforming was initially introduced to EEG analysis as a filter for source localization [26] to enhance the signal-to-noise ratio (SNR). Van Vliet et al. [20] first applied the spatiotemporal LCMV-beamformer as a method for the analysis of ERPs. The extension of this method to the combined spatiotemporal domain [20] and the data-driven approaches proposed by Treder et al. [27] and Wittevrongel et al. [21] allow for its application to classification problems.

For the following analyses, we assume that all EEG channels are normalized with zero mean and unit variance without loss of generality. Solving Equation (1) under the linear constraint given by Equation (2) returns the filter weights $\mathbf{w}$ defining the spatiotemporal LCMV-beamformer.

$$\arg\min_{\mathbf{w}} \mathbf{w}^\mathsf{T} C \mathbf{w}^\mathsf{T} \tag{1}$$

$$\mathbf{a}^\mathsf{T} \mathbf{w} = 1 \tag{2}$$

These weights minimize the variance of the output of the filter while enhancing the signal characterized by the constraint. $\mathbf{a} = \text{vec}(A)$ is the data-driven activation pattern, a template of the signal of interest maximizing the difference between two classes of epochs, determined as follows:

$$\mathbf{a} = \frac{1}{|\text{class 1}|} \sum_{\text{class 1}} \mathbf{x}_i - \frac{1}{|\text{class 2}|} \sum_{\text{class 2}} \mathbf{x}_i \quad (3)$$

The method of Lagrange multipliers then gives the closed-form solution to the minimization problem posed by Equations (1) and (2) as:

$$\mathbf{w} = \frac{C^{-1}\mathbf{a}^\top}{\mathbf{a}C^{-1}\mathbf{a}^\top} \quad (4)$$

The beamformer can be applied to epochs (unseen or not) as:

$$y_i = \mathbf{w}\mathbf{x}_i \quad (5)$$

resulting in a scalar output y_i per epoch. The linear constraint in Equation (2) ensures that the beamformer maps epochs containing a target response to a score close to one and, conversely, epochs not containing a target response to a score close to zero.

2.3. Covariance Matrix Regularization

Although the spatiotemporal beamformer, in theory, achieves optimal separation between target and non-target classes, in analogy to linear discriminant analysis [27], it does not always perform well on unseen data. The main challenge is to find a good estimator for the inverse covariance matrix C^{-1} since the real underlying covariance matrix generating the data is, in principle, unknown.

2.3.1. Empirical Covariance Estimation

Earlier spatiotemporal beamformer studies [21,22,28,29] use the empirical covariance and inverse covariance, calculated as follows:

$$\hat{C}_{\text{emp}} = \frac{1}{n-1} \sum_{i=1}^{n} \mathbf{x}_i \mathbf{x}_i^\top \quad (6)$$

$$\widehat{C^{-1}}_{\text{emp}} = \hat{C}_{\text{emp}}^{+} \quad (7)$$

The Moore–Penrose pseudoinverse $^+$ ensures that a solution exists when $\hat{C}_{\text{emp}}$ is singular. Figure 1a,b show examples of the empirical estimators of the covariance and the inverse covariance matrices, respectively. The empirical estimator suffers from performance and stability issues if the number of epochs n used for estimation is not much larger than the number of features cs [30,31].

2.3.2. Shrunk Covariance Estimation

The shrinkage covariance estimator creates a better-conditioned inversion matrix problem and generally performs better when applied to unseen data. The estimators for the covariance and inverse covariance are given by:

$$\hat{C}_\alpha = (1-\alpha)\hat{C}_{\text{emp}} + \alpha \frac{\text{Tr}(\hat{C}_{\text{emp}})}{cs} \mathbb{I} \quad (8)$$

$$\widehat{C^{-1}}_\alpha = \hat{C}_\alpha^{+} \quad (9)$$

with $0 < \alpha < 1$. Analogous to L2 regularization of the beamforming problem, shrinkage reduces the ratio between the smallest and largest eigenvalues of the covariance matrix by strengthening the diagonal. Figure 1c,d show examples of the shrunk estimator of the covariance and the inverse covariance matrices, respectively.

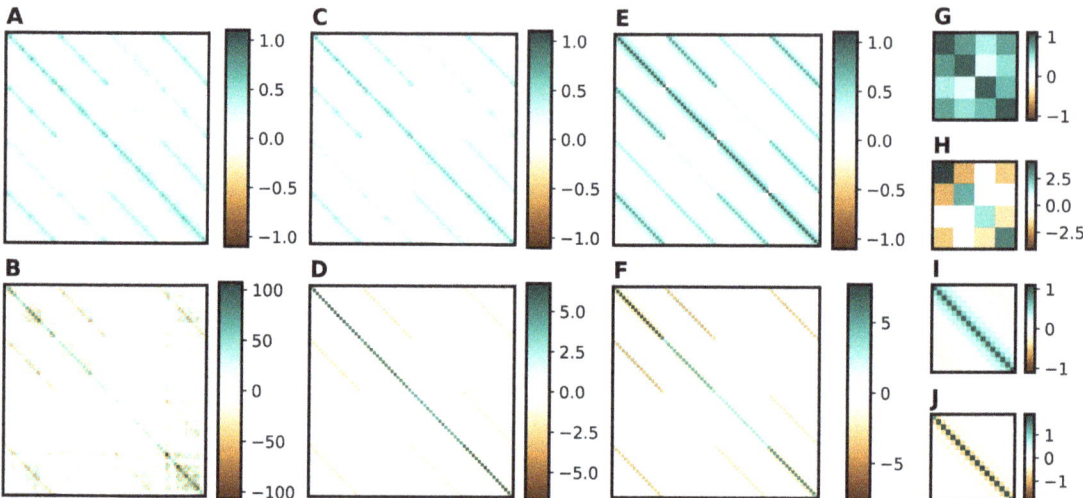

Figure 1. Different estimators of the covariance and inverse covariance of 100 epochs of data from *Subject 01* for channels *Fz*, *Cz*, *Pz*, and *Oz* and time samples between 0.1 s and 0.6 s. Regularized estimators of the inverse covariance exhibit less extreme values and have a sparser structure. (**A,B**) Empirical covariance and inverse covariance matrices. (**C,D**) Shrunk covariance and inverse covariance matrices with $\alpha = 0.14$ as determined by the closed-form leave-one-out cross-validation (LOOCV) method. (**E,F**) Kronecker–Toeplitz-structured covariance and inverse covariance matrices. (**G,H**) Spatial Kronecker factor of the Kronecker–Toeplitz-structured shrunk estimator and its inverse. (**I,J**) Temporal Kronecker factor of the Kronecker–Toeplitz-structured shrunk estimator and its inverse.

Earlier work [23] applied shrinkage regularization to ERP decoding with the spatiotemporal beamformer and showed competitive performance compared to other state-of-the-art decoding techniques such as stepwise LDA or SVM. The abovementioned researchers chose the shrinkage coefficient α as a fixed hyperparameter. However, its optimal value depends on the number of training epochs, the covariance matrix's dimensionality, and the independence and variance of the data, which can vary across evaluation settings and per session. The optimal value for α can be found with a line search using cross-validation method, but this can be a costly procedure. Methods exist to estimate an optimal shrinkage value directly from the data. Most notable among these are the Ledoit–Wolf procedure [32], the Rao–Blackwell Ledoit–Wolf method [33], and the oracle approximating shrinkage method [33]. A more recent estimation method [34] emulates a leave-one-out cross-validation (LOOCV) scheme expressed by the data-driven closed-form estimate:

$$\alpha = 1 - \frac{\frac{n}{n-1}\text{Tr}(\hat{C}_{\text{emp}}^2) - \frac{2}{cs}\left[\text{Tr}(\hat{C}_{\text{emp}})\right]^2 + \frac{1}{cs}\text{Tr}(\hat{C}_{\text{emp}}^2) - \frac{1}{n(n-1)}\sum_{i=1}^{n}||\mathbf{x}_i||_2^4}{\frac{n^2-2n}{(n-1)^2}\text{Tr}(\hat{C}_{\text{emp}}^2) - \frac{2}{cs}\left[\text{Tr}(\hat{C}_{\text{emp}})\right]^2 + \frac{1}{cs}\text{Tr}(\hat{C}_{\text{emp}}^2) + \frac{1}{n(n-1)^2}\sum_{i=1}^{n}||\mathbf{x}_i||_2^4} \quad (10)$$

Herein, we opt for the LOOCV shrinkage estimator because it avoids some of the assumptions made by [32,33] and because it generalizes to structured covariance estimations, as described in Section 2.3.3.

2.3.3. Spatiotemporal Beamforming with Kronecker–Toeplitz-Structured Covariance

Exploiting prior knowledge about the spatiotemporal structure of the EEG signal leads to a more regularized estimator of the covariance. When viewing the example of empirical spatiotemporal EEG covariance in Figure 1a, it becomes clear that this matrix consists of a block pattern of repeated, similar matrices. Due to the multi-channel nature of the signal, we assume that the covariance of spatiotemporal EEG epochs is a Kronecker product of two smaller matrices [35–37], as expressed by:

$$\hat{C}_{\text{struct}} = \hat{S} \otimes \hat{T} \tag{11}$$

with $\otimes$ denoting the Kronecker product operator. $\hat{S} \in \mathbb{R}^{c \times c}$ and $\hat{T} \in \mathbb{R}^{s \times s}$ correspond to estimators of the spatial and temporal covariance of the data, respectively. Furthermore, because the temporal covariance of the EEG-signal is stationary (i.e., it is only dependent on interval length between covarying time samples) [38], it is assumed to have a Toeplitz-matrix structure:

$$\hat{t}_{i,j} = \hat{t}_{i+1,j+1} \tag{12}$$

Property 1 then leads to Equation (13) to estimate the inverse covariance.

Property 1. $(U \otimes V)^+ = U^+ \otimes V^+$ *for any non-singular matrices U and V [39].*

$$\widehat{C^{-1}}_{\text{struct}} = \hat{S}^+ \otimes \hat{T}^+ \tag{13}$$

Finally, based on Property 2, Equation (4) can be reformulated more efficiently as Equation (14).

Property 2. $(U \otimes V) \cdot vec(W) = vec(VWU^\intercal)$ *for any matrices* $U \in \mathbb{R}^{p \times p}$, $V \in \mathbb{R}^{q \times q}$, *and* $W \in \mathbb{R}^{p \times q}$ *[40].*

$$\hat{w}_{\text{struct}} = \frac{\hat{S}^+ A^T \hat{T}^+}{\mathbf{a} \cdot vec(\hat{S}^+ A^T \hat{T}^+)} \tag{14}$$

Using Equation (14) removes the need to calculate the full, high-dimensional Kronecker product $\hat{S}^+ \otimes \hat{T}^+$. Figure 1e,f show examples of the structured covariance and inverse covariance estimators, respectively, consisting of a spatial Kronecker factor (Figure 1g,h) and a temporal component (Figure 1i,j).

The Kronecker approach has shown significant performance yields in different linear spatiotemporal EEG and MEG applications [24,37,41–43]. Van Vliet and Salmelin [25] applied a Kronecker-structured covariance estimator to ERP classification with linear models in a post hoc fashion. Our work goes further by embedding the Kronecker structure in the spatiotemporal beamformer training process, using a data-adaptive shrinkage method, and regularizing the covariance further by imposing a Toeplitz structure on the temporal covariance.

2.3.4. Kronecker–Toeplitz-Structured Covariance Estimation

The question remains how to estimate $\hat{S}$ and $\hat{T}$. Although the flip-flop and non-iterative flip-flop algorithms [44–46] can estimate Kronecker or Kronecker–Toeplitz-structured covariances, new results show that a fixed point iteration is more efficient [47,48]. After each iteration, the spatial and temporal covariance matrices are scaled to unit variance to ensure that the fixed-point iteration converges. Finally, shrinkage can also be introduced in the fixed-point iteration to improve stability and achieve more robust regularization [42,48–50].

The spatial and temporal covariance matrices are shrunk at every fixed-point iteration with shrinkage factors β_k and γ_k before matrix inversion in the next iteration. Combined, this leads to the iterative estimation algorithm described by the following equations:

$$\tilde{S}_{k+1} = \frac{1}{n} \sum_{i=1}^{n} X_i^\intercal \hat{T}_k^+ X_i \tag{15a}$$

$$\tilde{T}_{k+1} = \frac{1}{n}\sum_{i=1}^{n} X_i \hat{S}_k^+ X_i^\mathsf{T} \tag{15b}$$

$$\tilde{S}_{k+1}^{(\beta)} = (1 - \beta_{k+1})\tilde{S}_{k+1} + \beta_{k+1}\frac{\mathrm{Tr}(\tilde{S}_{k+1})}{c}\mathbb{I} \tag{16a}$$

$$\tilde{T}_{k+1}^{(\gamma)} = (1 - \gamma_{k+1})\tilde{T}_{k+1} + \gamma_{k+1}\frac{\mathrm{Tr}(\tilde{T}_{k+1})}{s}\mathbb{I} \tag{16b}$$

$$\hat{S}_{k+1} = \frac{c}{\mathrm{Tr}\left[\tilde{S}_{k+1}^{(\beta)}\right]}\tilde{S}_{k+1}^{(\beta)} \tag{17a}$$

$$\hat{T}_{k+1} = \frac{s}{\mathrm{Tr}\left[\tilde{T}_{k+1}^{(\gamma)}\right]}\tilde{T}_{k+1}^{(\gamma)} \tag{17b}$$

$\hat{S}_0$ and $\hat{T}_0$ can be initialized to any positive definite matrix. We choose to use the identity matrices $\mathbb{I}^{c\times c}$ and $\mathbb{I}^{s\times s}$. After each iteration, all diagonals of $\hat{R}_{k+1}$ are set to their mean values to ensure that $\hat{R}_{k+1}$ and $\hat{T}_{k+1}$ are Toeplitz-structured.

Xie et al. [51] show that the LOOCV estimates for the optimal values of β_{k+1} and γ_{k+1} also yield a closed-form solution for the Kronecker fixed-point-iteration algorithm:

$$\beta_{k+1} = 1 - \frac{\frac{n}{n-1}\mathrm{Tr}(\tilde{S}_{k+1}^2) - \frac{2}{c}\left[\mathrm{Tr}(\tilde{S}_{k+1})\right]^2 + \frac{1}{c}\mathrm{Tr}(\tilde{S}_{k+1}^2) - \frac{1}{n(n-1)}\sum_{i=1}^{n}\left[\mathrm{Tr}(X_i\hat{T}_k^+ X_i^\mathsf{T})^2\right]}{\frac{n^2-2n}{(n-1)^2}\mathrm{Tr}(\tilde{S}_{k+1}^2) - \frac{2}{c}\left[\mathrm{Tr}(\tilde{S}_{k+1})\right]^2 + \frac{1}{c}\mathrm{Tr}(\tilde{S}_{k+1}^2) + \frac{1}{(n-1)^2}\sum_{i=1}^{n}\left[\mathrm{Tr}(X_i\hat{T}_k^+ X_i^\mathsf{T})^2\right]} \tag{18a}$$

$$\gamma_{k+1} = 1 - \frac{\frac{n}{n-1}\mathrm{Tr}(\tilde{T}_{k+1}^2) - \frac{2}{s}\left[\mathrm{Tr}(\tilde{T}_{k+1})\right]^2 + \frac{1}{s}\mathrm{Tr}(\tilde{T}_{k+1}^2) - \frac{1}{n(n-1)}\sum_{i=1}^{n}\left[\mathrm{Tr}(X_i^\mathsf{T}\hat{S}_k^+ X_i)^2\right]}{\frac{n^2-2n}{(n-1)^2}\mathrm{Tr}(\tilde{T}_{k+1}^2) - \frac{2}{s}\left[\mathrm{Tr}(\tilde{T}_{k+1})\right]^2 + \frac{1}{s}\mathrm{Tr}(\tilde{T}_{k+1}^2) + \frac{1}{(n-1)^2}\sum_{i=1}^{n}\left[\mathrm{Tr}(X_i^\mathsf{T}\hat{S}_k^+ X_i)^2\right]} \tag{18b}$$

The shrinkage parameters $0 < \beta_{k+1} < 1$ and $0 < \gamma_{k+1} < 1$ should be re-determined after each iteration. The oracle approximation shrinkage method can also be used to determine β_{k+1} and γ_{k+1} [51,52] but performs worse for spatiotemporal EEG data since not all assumptions are met.

2.4. Dataset

We use the dataset from [21], containing P300 oddball EEG recordings of 21 healthy subjects since it is a high-quality dataset with a high number (32) of electrodes and concurrently recorded EOG responses for ocular artifact rejection. Nine targets were arranged on a monitor in front of the subject during an experimental session. The subject was asked to pay attention to a cued target for a block of stimulations. Within each block, the stimulations were organized in in 15 separate subsequent trials. A trial was defined as 9 stimulations in which each target is flashed precisely once per trial. Each target was cued four times, resulting in a dataset consisting of 36 blocks (4860 stimulations) per subject. Each stimulation corresponded to a single epoch in the preprocessed dataset. See [21] for a complete description of the dataset and the recording procedure.

2.5. Software and Preprocessing

Data processing and classifier analysis were performed in Python using Scikit-Learn (version 1.0.1) [53] and SciPy (version 1.7.1) [54]. The preprocessing pipeline was implemented using the MNE-Python toolbox (version 0.24.0) [55]. The dataset was converted to BIDS-EEG format [56] and managed and loaded with MNE-BIDS (version 0.9) [57]. The Riemannian classifier from Section 2.6.3 was implemented using pyRiemann (version 0.2.7). Statistical tests were performed in R (version 4.1.2).

The EEG recorded at 2048 Hz was re-referenced off-line to the average of the mastoids. The reference electrodes were dropped from the analysis. Data were subsequently filtered between 0.5Hz and 16Hz using forward-backward filtering with a fourth-order Butterworth IIR filter. The EEG signal was corrected for ocular artifacts using independent component analysis (ICA) by rejecting components that correlated with the bipolar EOG channels vEOG and hEOG, according to adaptive Z-score thresholding. Finally, epochs were cut from 0.1 s to 0.6 s after stimulus onset. No baseline correction was performed since this affects the temporal covariance of the data, violating the Toeplitz structure assumption [38].

2.6. Classification

2.6.1. Cross-Validation Scheme per Subject

We use a variation of grouped k-fold cross-validation per subject to evaluate the classifiers. We applied four-fold cross-validation by splitting the blocks of each subject into four continuous folds. Unlike regular cross-validation, we only used a single fold to train the classifiers while using the other three folds for validation. This scheme resulted in a training set of 9 blocks of 135 epochs each. We chose this approach since we are primarily interested in the performance of the classifiers in the case of low data availability. The classification task was to determine the cued target for each block. The fraction of correctly predicted cues provided the accuracy of a classifier. Data from all trials were used in the training stage, whereas classifier validation was performed multiple times per fold, each time using an increasing amount of trials (i.e., using the first trial, using the first two trials, etc., until all 15 trials have been used). For each of the 9 stimulated targets, the averages over the corresponding epochs across the utilized trials were used to predict the cued target in that block. The target with the maximum classifier score was then chosen as the predicted cued target. Before training the classifiers, a Z-score normalization transformation was developed on the training data to scale all EEG channels to unit variance. This transformation was then applied to the validation data.

2.6.2. Spatiotemporal Beamformer Classifier

Before calculating the spatiotemporal beamformer (STBF), the signal was downsampled to 32 Hz or twice the low-pass frequency 16 Hz, resulting in 17 time samples between 0.1 s and 0.6 s. According to the Nyquist theorem, more samples would not contain more information; hence, the minimum temporal resolution was chosen to reduce the dimensionality of the covariance. The activation pattern is the difference between the averages of epochs in response to cued targets and the averages of those in response to non-cued targets. We constructed three variations of the spatiotemporal beamformer: STBF with empirical covariance estimation (STBF-EMP) as in Section 2.3.1, STBF with LOOCV-shrunk covariance estimation (STBF-SHRUNK) as in Section 2.3.2, and STBF with Kronecker–Toeplitz-structured covariance estimation (STBF-STRUCT) with LOOCV shrinkage for the Kronecker factors as in Section 2.3.4.

2.6.3. Riemannian Geometry Classifier

We opted for a Riemannian geometry-based classifier to compare our results. The Riemannian model (xDAWN+RG) uses the xDAWN spatial filter combined with Riemannian geometry in tangent space as implemented by Barachant et al. [58]. This classifier uses four xDAWN spatial filters and each epoch's empirical spatial covariance matrix. The target with the maximum score is the prediction of the cued target. xDAWN+RG was trained and validated without downsampling using epochs at the original sample rate of 2048 Hz.

3. Results

3.1. Minimum Required Fixed-Point Iterations

The fixed point iteration algorithm described in Equations (15a)–(16b) is used to estimate the Kronecker–Toeplitz-structured covariance for the STBF-STRUCT classifier. Fixed-point iteration is an iterative procedure starting from (in our case) non-informed initial guesses for the spatial and temporal covariance matrices. As a stopping criterion, one could impose a threshold on the difference in outcome of successive steps, e.g., based on the covariance norm or the classifier accuracy. However, few iterations or even just one [59] suffice to achieve satisfactory performance in practice.

Figure 2 confirms these results for the STBF-STRUCT classifier. Using more than one fixed-point iteration does not significantly improve the accuracy across the amounts of training data and the number of trials used for evaluation. Hence, only one iteration is used for the STBF-STRUCT classifier, leading to a drastic speed-up of the training process.

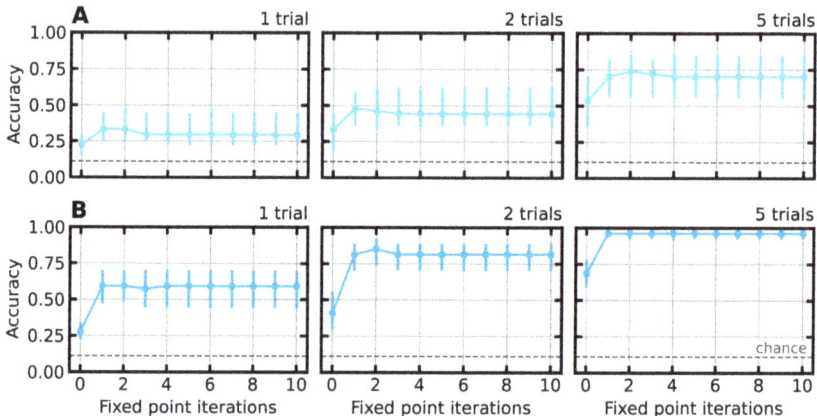

Figure 2. Average cross-validated STBF-STRUCT accuracy using one trial per block for validation over all 21 subjects relative to the number of iterations used to estimate the Kronecker–Toeplitz-structured shrunk covariance. Error bars represent the first and third quartiles. The accuracy does not improve when using more than one iteration. (**A**) Results for 1, 2, and 5 trials using only the first block in each training fold for training. (**B**) Results for 1, 2, and 5 trials using all nine training blocks in the training folds.

3.2. Classifier Accuracy for Limited Training Data

It is of interest to keep the calibration time before BCI operation as short as possible. We mimic this problem by training the classifier with as few training epochs as possible. We evaluate the performance of all classifiers for different levels of available training data and apply the cross-validation procedure nine times (the number of blocks in the training fold) for all subjects, keeping the corresponding number of blocks in the training folds and dropping the rest. Figures 3 and A1 show each classifier's accuracy relative to the data availability. We statistically compare the two newly proposed classifiers, STBF-STRUCT and STBF-SHRUNK, for different levels of training data availability using a one-sided paired Wilcoxon rank-sum test with Holm correction for the multiple pairwise comparisons between classifiers. We performed this analysis three times: by only using the first trial of a block, by averaging epochs across the first two trials of a block, and across the first five trials of a block. Results validated on one trial are reported in Table 1, two-trial results in Table 2, and five-trial results in Table 3.

Table 1. p-values calculated via the one-sided paired Wilcoxon rank-sum test with Holm correction using one testing trial for different classifiers and levels of data availability. p-values < 0.05 are considered significant and marked bold.

1 Trial	Nb. of Training Blocks								
	1	2	3	4	5	6	7	8	9
STBF-STRUCT > STBF-SHRUNK	0.005	0.030	0.015	0.543	0.284	0.159	–	–	0.952
STBF-STRUCT > STBF-EMP	<0.001	<0.001	<0.001	<0.001	<0.001	<0.001	<0.001	<0.001	<0.001
STBF-STRUCT > xDAWN+RG	0.086	0.002	<0.001	<0.001	<0.001	<0.001	<0.001	<0.001	<0.001
STBF-SHRUNK > STBF-EMP	<0.001	<0.001	<0.001	<0.001	<0.001	<0.001	<0.001	<0.001	<0.001
STBF-SHRUNK > xDAWN+RG	–	0.499	0.071	<0.001	<0.001	<0.001	<0.001	0.001	0.001

Table 2. *p*-values as in Table 1, averaging over two testing trials.

2 Trials	Nb. of Training Blocks								
	1	2	3	4	5	6	7	8	9
STBF-STRUCT > STBF-SHRUNK	0.014	0.006	0.040	0.040	0.004	0.846	0.888	–	–
STBF-STRUCT > STBF-EMP	<0.001	<0.001	<0.001	<0.001	<0.001	<0.001	<0.001	<0.001	<0.001
STBF-STRUCT > xDAWN+RG	0.103	0.004	<0.001	<0.001	<0.001	<0.001	<0.001	<0.001	<0.001
STBF-SHRUNK > STBF-EMP	<0.001	<0.001	<0.001	<0.001	<0.001	<0.001	<0.001	<0.001	<0.001
STBF-SHRUNK > xDAWN+RG	–	–	0.163	0.001	0.001	<0.001	<0.001	<0.001	<0.001

Table 3. *p*-values as in Table 1, averaging over five testing trials.

5 Trials	Nb. of Training Blocks								
	1	2	3	4	5	6	7	8	9
STBF-STRUCT > STBF-SHRUNK	0.005	0.030	0.015	0.543	0.284	0.159	–	–	0.952
STBF-STRUCT > STBF-EMP	<0.001	<0.001	<0.001	<0.001	<0.001	<0.001	<0.001	<0.001	<0.001
STBF-STRUCT > xDAWN+RG	0.086	0.002	<0.001	<0.001	<0.001	0.004	0.006	<0.001	<0.001
STBF-SHRUNK > STBF-EMP	<0.001	<0.001	<0.001	<0.001	<0.001	<0.001	<0.001	<0.001	<0.001
STBF-SHRUNK > xDAWN+RG	–	0.499	<0.001	<0.001	<0.001	<0.001	<0.001	0.001	0.001

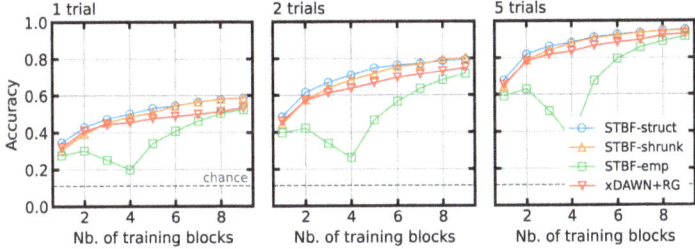

Figure 3. Accuracy of the different classifiers for all 21 subjects relative to the number of blocks available for training. One block consists of 135 epochs and corresponds to 27 seconds of stimulation. Accuracies are shown for the evaluation settings averaging over 1, 2, and 3 trials of testing stimuli. Figure A1 contains results for all numbers of trials. Although STBF-EMP is unstable when few training data are available, regularization of the covariance matrix (STBF-SHRUNK and STBF-STRUCT) drastically improves performance.

The tables show that STBF-STRUCT has a significant advantage over STBF-SHRUNK when the number of training blocks is low. This effect is present for 1-, 2-, and 5-trial evaluations. This advantage decreases (the *p*-value increases) when adding more training blocks. Both STBF-STRUCT and STBF-SHRUNK perform significantly better than STBF-EMP for all evaluated settings. Compared to xDAWN+RG, STBF-STRUCT also has significantly higher accuracy in almost all evaluated settings, except when using only one training block. STBF-SHRUNK does not outperform xDAWN+RG when training data are scarce but gains a significant advantage when using more training data.

3.3. Classifier Training Time

In order to evaluate the training time of the investigated classifiers, the cross-validation scheme was run four times for each subject, each time with an increasing number of EEG channels retained in the analysis, to explore the scalability of each classifier for analyses with higher spatial resolutions. The temporal resolutions were not varied, but we expect that increasing the temporal resolution has a similar effect on training time, since the training times for the STBF-based classifiers are primarily dependent on the number of parameters in their respective covariance matrix estimators, as evidenced by the complexity calculations in Section 4.2. Figure 4 shows the measured training times. These results were obtained using a laptop with an Intel® Core™ i7-8750H CPU (Intel Corporation, Santa Clara, CA, USA) and 16 GB of RAM.

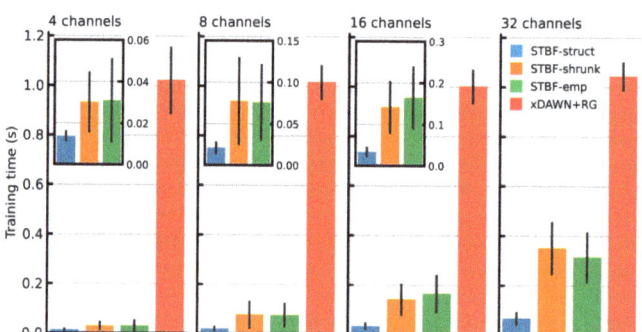

Figure 4. Median fold training time for different classifiers at different spatial resolution levels evaluated over all training folds for all subjects. Error bars represent standard deviation. The training time of STBF-SHRUNK increased more steeply with resolution compared to STBF-STRUCT. All STBF classifiers were able to be trained significantly faster than xDAWN+RG.

Figure 4 shows that the training time of STBF-STRUCT increased less steeply than that of STBF-SHRUNK and STBF-EMP. The training time of all three STBF-based classifiers was much lower than that of xDAWN+RG, which appears nearly constant when using 4, 8, 16, or 32 channels.

4. Discussion

4.1. Classification Accuracy

As evidenced by Figure 3 and Tables 1–3, the regularized classifiers STBF-SHRUNK and STBF-STRUCT significantly improve the classification accuracy compared to the original STBF-EMP for all the numbers of training blocks indicated. We believe there are three reasons for this. First and foremost, the empirical covariance matrix in STBF-EMP becomes ill-conditioned when the number of available training epochs is smaller than the number of features ($n < cs$), rendering its inversion with the Moore–Penrose pseudoinverse unstable. This is the case for STBF-EMP when $n = cs = 32 \times 17 = 544$, after which the accuracy of STBF-EMP starts to increase. This effect is visible in Figure 3, where the accuracy starts increasing when using more than four training blocks, amounting to 540 epochs. The noticeable dip in accuracy when using around 540 epochs can be explained by numerical effects in the pseudoinverse for very small eigenvalues [60–63]. Regularization of the covariance matrix with shrinkage ensures that the covariance matrix is non-singular and better conditioned so that it can stably be inverted. Second, covariance regularization introduces a trade-off between the variance and bias of the model [32]. Better performance on unseen data can be achieved when some model variance is traded for extra bias. Regularization reduces the extreme values present, as shown in Figure 1, resulting in a classifier with better generalization. Third, the true spatiotemporal covariance matrix may vary throughout BCI sessions, e.g., due to movement of the EEG-cap, changing impedances of electrodes, subject fatigue, the introduction of new spatiotemporal noise sources, and other possible confounds. A regularized covariance matrix should better account for changes in true covariance. Note that the LOOCV method in principle assumes that the covariances of the training data and unseen data are the same. Because the covariance might have changed for unseen data, the shrinkage estimate obtained with LOOCV is probably still an underestimation of the optimal—but unknown—shrinkage coefficient that would yield the best classification accuracy for the unseen data.

Another observation is the significantly better accuracy score of STBF-STRUCT over STBF-SHRUNK when the amount of available training data is small. This property is an attractive advantage in a BCI setting since it is desirable to keep the calibration (training) phase as short as possible without losing accuracy. The accuracy advantage of the structured estimator is a consequence of the Kronecker–Toeplitz covariance structure, which is informative for the underlying process generating the epochs, if it is assumed that the EEG signal is a linear combination of stationary activity generated by random dipoles in the brain with added noise [24,35,41]. Hence, STBF-STRUCT can utilize this prior information to better estimate the inverse covariance. The increase in accuracy for small training set sizes can also be explained by the smaller number of parameters necessary to estimate the inverse covariance (see Section 4.2), increasing the stability of matrix inversions.

When compared to the state-of-the-art xDAWN+RG classifier, we conclude that STBF-STRUCT reaches similar accuracy when using only one block of training data. The authors suspect this is due to both classifiers having insufficient training information to reach satisfactory classification accuracy. When more data are available, STBF-STRUCT reaches significantly better accuracies. Combined with the benefits laid out in Sections 4.2 and 4.3, this makes it an attractive option for ERP classification. STBF-SHRUNK does not show decisive accuracy improvements over xDAWN+RG using a few training blocks, but this improves as the training data increases.

4.2. Time and Memory Complexity

As mentioned above, inverting the full $cs \times cs$ dimensional covariance matrix to construct STBF-EMP and STBF-SHRUNK can be costly and unstable, in particular in high-resolution settings with many EEG channels or time samples. Constructing the full covariance and inverse covariance matrices also requires a considerable amount of memory. The structured covariance estimator of STBF-STRUCT has two advantages here.

First, because of Properties 1 and 2 there is no need to calculate the full $cs \times cs$ symmetric covariance and inverse covariance matrices for STBF-STRUCT or keep them in memory; they can instead be replaced by two smaller symmetric matrices of dimensions $c \times c$ and $s \times s$, respectively. Furthermore, since the temporal component of the Kronecker product is Toeplitz-structured, it only requires s parameters to estimate. Although the inverse covariance of STBF-EMP and STBF-SHRUNK is defined by $\frac{cs(cs+1)}{2} = \frac{32 \times 17(32 \times 17+1)}{2} = 122.128$ parameters accounting for the symmetric nature of covariance, the structured estimator only requires $\frac{c(c+1)}{2} + s = \frac{32(32+1)}{2} + 17 = 545$ unique parameters. This reduction in parameters to estimate reduces memory usage and contributes to the regularization effect for low-data-availability settings. The inverse covariances of STBF-EMP and STBF-STRUCT, represented as $32 \times 17 \times 32 \times 17$ symmetric matrices of single-precision real floating point numbers for weight calculation, use 9.03 MiB of memory. The 32×32 and 17×17 matrices of STBF-STRUCT only require 5.12 KiB.

Second, structured estimation has better time complexity. Covariance estimation and inversion occupy the largest part of the STBF training time. For STBF-EMP and STBF-SHRUNK, the time complexity of this process is $\mathcal{O}(nc^2s^2 + c^3s^3)$. Thanks to Property 1, the complexity can be reduced to $\mathcal{O}(nc^2s^2 + c^3 + s^3)$ for the structured estimator of STBF-STRUCT. The results presented in Figure 4 confirm these calculations. It can be observed that the training time of STBF-STRUCT stays low compared to STBF-EMP and STBF-SHRUNK when dimensionality increases.

The results shown in Figure 4 also confirm that the STBF-based estimators are very fast to train compared to the state-of-the-art estimator xDAWN+RG, which confirms the results in [21]. Since the training times of all STBF-based classifiers are already of the order of tenths of seconds, the question arises as to whether the improvements achieved using the structured estimator would be relevant in practice. However, the authors believe that these results could significantly impact some use-cases of the spatiotemporal beamformer, such as high-spatial- or temporal-resolution ERP analyses. One example is single-trial ERP analysis with a high temporal resolution to extract ERP time features. Such higher-resolution analyses can later be incorporated into an ERP classification framework. In addition, the speed provided by structured estimation yields a faster off-line evaluation of the STBF ERP classifier, in which multiple cross-validation folds, subjects, and hyperparameter settings often need to be explored, which can quickly increase runtime. Improvements in computation speed and memory usage can remove the need for dedicated computation hardware and enable group analyses to be run on a personal computer.

4.3. Interpreting the Weights

The weight matrix of the STBF determines how each spatiotemporal feature of a given epoch should contribute to enhancing the SNR of the discriminating signal in the classification task. Alternatively, the activation pattern can be regarded as a forward EEG model of the activity, generating the discriminating signal and the weights as a backward model [60,64]. Regularization enables a researcher to interpret better the distribution of the weight over space and time after reshaping the weight vector **w** to its spatiotemporal matrix equivalent, W, such that $\text{vec}(W) = \mathbf{w}$. Figure 5 compares the weights calculated in STBF-EMP and STBF-SHRUNK with the weights from STBF-STRUCT.

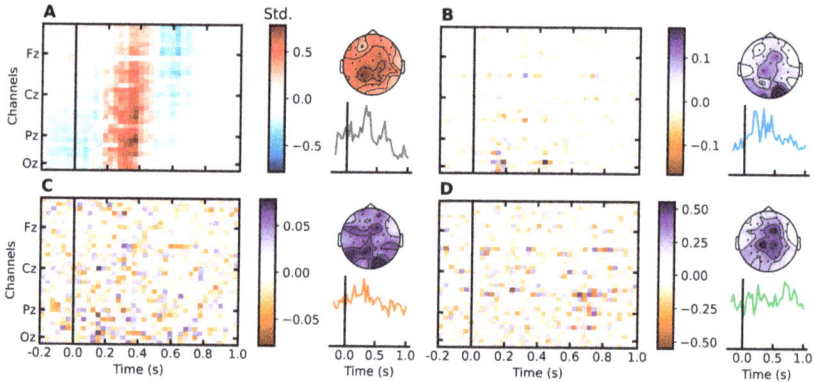

Figure 5. Spatiotemporal beamformer weights calculated using four blocks of data (of 1215 epochs) from *Subject 01* from 0.2 s before until 1.0 s after stimulus onset. Regularized weights show an interpretable sparse pattern, whereas the empirical weights appear noisier. (**A**) Spatiotemporal activation pattern with spatial and temporal global field power. (**B**) STBF-STRUCT weights with spatial and temporal averages of absolute values. (**C**) STBF-SHRUNK weights. The shrinkage factor $\alpha = 0.05$ was determined with the closed-form LOOCV method. (**D**) STBF-EMP weights.

Since the linear filter's noise suppression and signal amplification functions are deeply entangled, it is not necessarily true that features with a high filter weight directly correlate to features containing discriminatory information [64]. However, it is still possible to interpret the weights in terms of which features contribute most to the classification process, be it through noise suppression, signal amplification, or—most likely—a combination of both. The weights obtained by STBF-EMP seem to be randomly distributed over space and time; the regularized estimator used by STBF-SHRUNK and STBF-STRUCT reveal a more interpretable weight distribution. The STBF-SHRUNK weights show a sparse spatial distribution, whereas the STBF-STRUCT weights show a sparse distribution in both space and in time.

As expected, Figure 5b and d exhibit weight around the central and parietal regions, where the P300 ERP component is present. Especially the spatial weights of STBF-SHRUNK in Figure 5d correspond to the spatial activation pattern in Figure 5a. This is not surprising, since shrinkage transforms the covariance matrix closer to the identity matrix and assuming identity covariance in Figure 4 yields weights identical to the activation pattern (up to a scaling factor). Additionally, Figure 5b shows that weights in the baseline interval and after 0.6 s, which should contain no response information, are close to zero for the structured estimator. Meanwhile, these weights are high in the occipital region between 0.1 s and 0.2 s, containing early visual components with relatively low SNR. This high weight for the early visual components confirms the results from Treder and Blankertz [65] that state that, in addition to the P300, the early N1 and P2 ERP components are also modulated by oddball attention and contain discriminatory information between attended and non-attended stimuli.

Using an interpretable classification model has many advantages. For instance, one can use the weight matrix to determine relevant time samples and EEG channels for per-subject feature selection to refine the model further. The number of channels is also an important cost factor in practical BCI applications. Determining which channels do not contribute to the classification accuracy helps to reduce the number of required electrodes. Spatially clustered weights indicate that some electrodes are not used by the classifier and can be discarded accordingly with no substantial accuracy reduction. As another example, information about the timing and spatial distribution of the discriminatory information in the response can be extracted from the weights and linked to neurophysiological hypotheses.

5. Conclusions

Although it is possible to regularize the spatiotemporal LCMV beamformer classifier for ERP detection through other methods, such as by employing feature selection, by adding regularizing penalties to the cost function beamforming problem, or by crafting a cleaner activation pattern, our work focused on estimation methods for spatiotemporal covariance. We introduced a covariance estimator using adaptive shrinkage and an estimator exploiting prior knowledge about the spatiotemporal nature of the EEG signal. We compared these estimators with the original spatiotemporal beamformer and a state-of-the-art method in an off-line P300 detection task. Our results show that the structured estimator performs better when training data are sparsely available and that results can be computed faster and with substantially less memory usage. Since these algorithms are not paradigm-specific, the conclusions can be generalized to other ERP-based BCI settings.

Future work should focus on introducing more robust regularization strategies using prior knowledge, such as shrinking the spatial covariance to a population mean or a previously known matrix based on sensor geometry or characterizing the temporal covariance as a wavelet or autoregressive model. More accurate results could be obtained by expressing the covariance as the sum of multiple Kronecker products to account for spatial variation in temporal covariance. It could also be interesting to explore the impact of covariance regularization on transfer learning of the STBF between subjects to alleviate calibration entirely. Finally, it could be insightful to evaluate the proposed algorithms in a real-world on-line BCI setting.

Author Contributions: Conceptualization, A.V.D.K.; methodology, A.V.D.K., A.L., B.W. and M.M.V.H.; software, A.V.D.K., validation, A.V.D.K.; formal analysis, A.V.D.K. and B.W.; investigation, A.V.D.K. and B.W.; resources, M.M.V.H. and B.W.; data curation, A.V.D.K.; writing—original draft preparation, A.V.D.K.; writing—review and editing, A.V.D.K., A.L., B.W. and M.M.V.H.; visualization, A.V.D.K.; supervision, M.M.V.H.; project administration, A.V.D.K. and M.M.V.H.; funding acquisition, M.M.V.H. All authors have read and agreed to the published version of the manuscript.

Funding: A.V.D.K. is supported by the special research fund of the KU Leuven (GPUDL/20/031). A.L. is supported by the Belgian Fund for Scientific Research—Flanders (1SC3419N). M.M.V.H. is supported by research grants received from the European Union's Horizon 2020 research and innovation programme under grant agreement No. 857375, the special research fund of the KU Leuven (C24/18/098), the Belgian Fund for Scientific Research—Flanders (G0A4118N, G0A4321N, G0C1522N), and the Hercules Foundation (AKUL 043).

Institutional Review Board Statement: The study was conducted according to the guidelines of the Declaration of Helsinki, and approved by the Ethics Committee of KU Leuven's university hospital (UZ Leuven) (S62547 approved 11 June 2019).

Informed Consent Statement: Informed consent was obtained from all subjects involved in the study.

Data Availability Statement: No new data were created or analyzed in this study. Data sharing is not applicable to this article. Source code is available at https://github.com/kul-compneuro/stbf-erp (accessed 7 March 2022).

Acknowledgments: The authors acknowledge François Cabestaing and Hakim Si-Mohammed for their valuable input in the development of this article.

Conflicts of Interest: The authors declare no conflict of interest. The funders had no role in the design of the study; in the collection, analyses, or the interpretation of data; in the writing of the manuscript, or in the decision to publish the results.

Appendix A

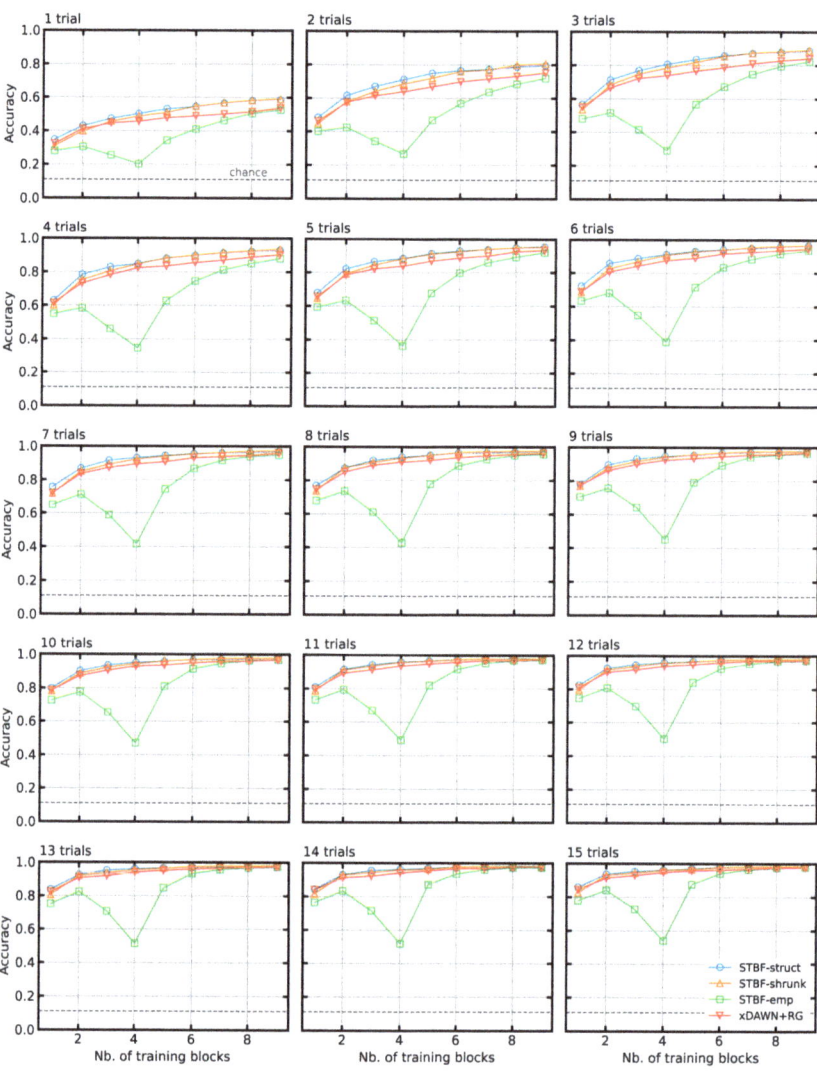

Figure A1. Accuracy of the different classifiers for all 21 subjects relative to the number of blocks available for training. One block consists of 135 epochs and corresponds to 27 s of stimulation. Accuracies are shown for the evaluation settings, averaging over different numbers of trials, ranging from 1 to 15.

References

1. Wolpaw, J.R.; Birbaumer, N.; McFarland, D.J.; Pfurtscheller, G.; Vaughan, T.M. Brain–computer interfaces for communication and control. *Clin. Neurophysiol.* **2002**, *113*, 767–791. [CrossRef]
2. Naci, L.; Monti, M.M.; Cruse, D.; Kübler, A.; Sorger, B.; Goebel, R.; Kotchoubey, B.; Owen, A.M. Brain–computer interfaces for communication with nonresponsive patients. *Ann. Neurol.* **2012**, *72*, 312–323. [CrossRef] [PubMed]
3. Chaudhary, U.; Birbaumer, N.; Ramos-Murguialday, A. Brain-computer interfaces for communication and rehabilitation. *Nat. Rev. Neurol.* **2016**, *12*, 513–525. [CrossRef] [PubMed]
4. Abiri, R.; Borhani, S.; Sellers, E.W.; Jiang, Y.; Zhao, X. A comprehensive review of EEG-based brain–computer interface paradigms. *J. Neural Eng.* **2019**, *16*, 11001. [CrossRef] [PubMed]
5. Mellinger, J.; Schalk, G.; Braun, C.; Preissl, H.; Rosenstiel, W.; Birbaumer, N.; Kübler, A. An MEG-based brain–computer interface (BCI). *NeuroImage* **2007**, *36*, 581–593. [CrossRef] [PubMed]
6. Hong, K.S.; Naseer, N.; Kim, Y.H. Classification of prefrontal and motor cortex signals for three-class fNIRS–BCI. *Neurosci. Lett.* **2015**, *587*, 87–92. [CrossRef] [PubMed]
7. Paek, A.Y.; Kilicarslan, A.; Korenko, B.; Gerginov, V.; Knappe, S.; Contreras-Vidal, J.L. Towards a Portable Magnetoencephalography Based Brain Computer Interface with Optically-Pumped Magnetometers. In Proceedings of the 2020 42nd Annual International Conference of the IEEE Engineering in Medicine Biology Society (EMBC), Montreal, QC, Canada, 20–24 July 2020; pp. 3420–3423. [CrossRef]
8. Schalk, G.; Leuthardt, E.C. Brain-computer interfaces using electrocorticographic signals. *IEEE Rev. Biomed. Eng.* **2011**, *4*, 140–154. [CrossRef] [PubMed]
9. Maynard, E.M.; Nordhausen, C.T.; Normann, R.A. The Utah Intracortical Electrode Array: A recording structure for potential brain-computer interfaces. *Electroencephalogr. Clin. Neurophysiol.* **1997**, *102*, 228–239. [CrossRef]
10. Willett, F.R.; Avansino, D.T.; Hochberg, L.R.; Henderson, J.M.; Shenoy, K.V. High-performance brain-to-text communication via handwriting. *Nature* **2021**, *593*, 249–254. [CrossRef]
11. Gao, S.; Wang, Y.; Gao, X.; Hong, B. Visual and Auditory Brain–Computer Interfaces. *IEEE Trans. Biomed. Eng.* **2014**, *61*, 1436–1447. [CrossRef]
12. Kapgate, D.; Kalbande, D. A Review on Visual Brain Computer Interface. In *Advancements of Medical Electronics*; Lecture Notes in Bioengineering; Gupta, S., Bag, S., Ganguly, K., Sarkar, I., Biswas, P., Eds.; Springer India: New Delhi, India, 2015; pp. 193–206. [CrossRef]
13. Farwell, L.A.; Donchin, E. Talking off the top of your head: Toward a mental prosthesis utilizing event-related brain potentials. *Electroencephalogr. Clin. Neurophysiol.* **1988**, *70*, 510–523. [CrossRef]
14. Sellers, E.W.; Donchin, E. A P300-based brain–computer interface: Initial tests by ALS patients. *Clin. Neurophysiol.* **2006**, *117*, 538–548. [CrossRef] [PubMed]
15. Barachant, A.; Congedo, M. A Plug&Play P300 BCI Using Information Geometry. *arXiv* **2014**, arXiv:1409.0107.
16. Philip, J.T.; George, S.T. Visual P300 Mind-Speller Brain-Computer Interfaces: A Walk through the Recent Developments with Special Focus on Classification Algorithms. *Clin. EEG Neurosci.* **2020**, *51*, 19–33. [CrossRef] [PubMed]
17. Tayeb, S.; Mahmoudi, A.; Regragui, F.; Himmi, M.M. Efficient detection of P300 using Kernel PCA and support vector machine. In Proceedings of the 2014 Second World Conference on Complex Systems (WCCS), Agadir, Morocco, 10–12 November 2014; pp. 17–22. [CrossRef]
18. Vařeka, L. Evaluation of convolutional neural networks using a large multi-subject P300 dataset. *Biomed. Signal Process. Control* **2020**, *58*, 101837. [CrossRef]
19. Borra, D.; Fantozzi, S.; Magosso, E. Convolutional neural network for a P300 brain-computer interface to improve social attention in autistic spectrum disorder. In *Mediterranean Conference on Medical and Biological Engineering and Computing—MEDICON 2019*; Henriques, J., Neves, N., de Carvalho, P., Eds.; Springer International Publishing: Cham, Switzerland, 2020; pp. 1837–1843. [CrossRef]
20. Van Vliet, M.; Chumerin, N.; De Deyne, S.; Wiersema, J.R.; Fias, W.; Storms, G.; Van Hulle, M.M. Single-trial erp component analysis using a spatiotemporal lcmv beamformer. *IEEE Trans. Biomed. Eng.* **2015**, *63*, 55–66. [CrossRef] [PubMed]
21. Wittevrongel, B.; Van Hulle, M.M. Faster p300 classifier training using spatiotemporal beamforming. *Int. J. Neural Syst.* **2016**, *26*, 1650014. [CrossRef]
22. Wittevrongel, B.; Van Hulle, M.M. Spatiotemporal Beamforming: A Transparent and Unified Decoding Approach to Synchronous Visual Brain-Computer Interfacing. *Front. Neurosci.* **2017**, *11*, 630. [CrossRef]
23. Libert, A.; Wittevrongel, B.; Van Hulle, M.M. Effect of stimulus direction on motion-onset visual evoked potentials decoded using spatiotemporal beamforming Abstract. In Proceedings of the 2021 10th International IEEE/EMBS Conference on Neural Engineering (NER), Virtual, 4–6 May 2021; pp. 503–506. [CrossRef]
24. Gonzalez-Navarro, P.; Moghadamfalahi, M.; Akçakaya, M.; Erdogmus, D. Spatio-temporal EEG models for brain interfaces. *Signal Process.* **2017**, *131*, 333–343. [CrossRef] [PubMed]
25. Van Vliet, M.; Salmelin, R. Post-hoc modification of linear models: Combining machine learning with domain information to make solid inferences from noisy data. *NeuroImage* **2020**, *204*, 116221. [CrossRef]
26. Van Veen, B.D.; Van Drongelen, W.; Yuchtman, M.; Suzuki, A. Localization of brain electrical activity via linearly constrained minimum variance spatial filtering. *IEEE Trans. Biomed. Eng.* **1997**, *44*, 867–880. [CrossRef] [PubMed]

27. Treder, M.S.; Porbadnigk, A.K.; Avarvand, F.S.; Müller, K.R.; Blankertz, B. The LDA beamformer: Optimal estimation of ERP source time series using linear discriminant analysis. *NeuroImage* **2016**, *129*, 279–291. [CrossRef] [PubMed]
28. Wittevrongel, B.; Hulle, M.M.V. Frequency- and Phase Encoded SSVEP Using Spatiotemporal Beamforming. *PLoS ONE* **2016**, *11*, e0159988. [CrossRef] [PubMed]
29. Wittevrongel, B.; Van Wolputte, E.; Van Hulle, M.M. Code-modulated visual evoked potentials using fast stimulus presentation and spatiotemporal beamformer decoding. *Sci. Rep.* **2017**, *7*, 15037. [CrossRef] [PubMed]
30. Stein, C. Inadmissability of the usual estimator for the mean of a multivariate normal distribution. In *Contribution to the Theory of Statistics*; University of California Press: Berkeley, CA, USA, 1956; Volume 1, pp. 197–206. [CrossRef]
31. Khatri, C.; Rao, C. Effects of estimated noise covariance matrix in optimal signal detection. *IEEE Trans. Acoust. Speech, Signal Process.* **1987**, *35*, 671–679. [CrossRef]
32. Ledoit, O.; Wolf, M. A well-conditioned estimator for large-dimensional covariance matrices. *J. Multivar. Anal.* **2004**, *88*, 365–411. [CrossRef]
33. Chen, Y.; Wiesel, A.; Eldar, Y.C.; Hero, A.O. Shrinkage algorithms for MMSE covariance estimation. *IEEE Trans. Signal Process.* **2010**, *58*, 5016–5029. [CrossRef]
34. Tong, J.; Hu, R.; Xi, J.; Xiao, Z.; Guo, Q.; Yu, Y. Linear shrinkage estimation of covariance matrices using low-complexity cross-validation. *Signal Process.* **2018**, *148*, 223–233. [CrossRef]
35. De Munck, J.; Vijn, P.; Lopes da Silva, F. A random dipole model for spontaneous brain activity. *IEEE Trans. Biomed. Eng.* **1992**, *39*, 791–804. [CrossRef]
36. De Munck, J.C.; Van Dijk, B.W. *The Spatial Distribution of Spontaneous EEG and MEG*; Springer Series in Synergetics; Springer: Berlin/Heidelberg, Germany, 1999; pp. 202–228. [CrossRef]
37. Huizenga, H.M.; De Munck, J.C.; Waldorp, L.J.; Grasman, R.P. Spatiotemporal EEG/MEG source analysis based on a parametric noise covariance model. *IEEE Trans. Biomed. Eng.* **2002**, *49*, 533–539. [CrossRef]
38. Bijma, F.; de Munck, J.C.; Huizenga, H.M.; Heethaar, R.M. A mathematical approach to the temporal stationarity of background noise in MEG/EEG measurements. *NeuroImage* **2003**, *20*, 233–243. [CrossRef]
39. Langville, A.N.; Stewart, W.J. The Kronecker product and stochastic automata networks. *J. Comput. Appl. Math.* **2004**, *167*, 429–447. [CrossRef]
40. Loan, C.F.V. The ubiquitous Kronecker product. *J. Comput. Appl. Math.* **2000**, *123*, 85–100. [CrossRef]
41. De Munck, J.C.; Huizenga, H.M.; Waldorp, L.J.; Heethaar, R. Estimating stationary dipoles from MEG/EEG data contaminated with spatially and temporally correlated background noise. *IEEE Trans. Signal Process.* **2002**, *50*, 1565–1572. [CrossRef]
42. Beltrachini, L.; von Ellenrieder, N.; Muravchik, C.H. Shrinkage Approach for Spatiotemporal EEG Covariance Matrix Estimation. *IEEE Trans. Signal Process.* **2013**, *61*, 1797–1808. [CrossRef]
43. Gonzalez-Navarro, P.; Moghadamfalahi, M.; Akcakaya, M.; Erdogmus, D. A kronecker product structured EEG covariance estimator for a language model assisted-BCI. In *International Conference on Augmented Cognition*; Springer: Cham, Switzerland, 2016; pp. 35–45. [CrossRef]
44. Lu, N.; Zimmerman, D.L. The likelihood ratio test for a separable covariance matrix. *Stat. Probab. Lett.* **2005**, *73*, 449–457. [CrossRef]
45. Werner, K.; Jansson, M.; Stoica, P. On estimation of covariance matrices with Kronecker product structure. *IEEE Trans. Signal Process.* **2008**, *56*, 478–491. [CrossRef]
46. Wirfält, P.; Jansson, M. On Toeplitz and Kronecker structured covariance matrix estimation. In Proceedings of the 2010 IEEE Sensor Array and Multichannel Signal Processing Workshop, Jerusalem, Israel, 4–7 October 2010; pp. 185–188. [CrossRef]
47. Wiesel, A. Geodesic Convexity and Covariance Estimation. *IEEE Trans. Signal Process.* **2012**, *60*, 6182–6189. [CrossRef]
48. Wiesel, A. On the convexity in Kronecker structured covariance estimation. In Proceedings of the 2012 IEEE Statistical Signal Processing Workshop (SSP), Ann Arbor, MI, USA, 5–8 August 2012; pp. 880–883. [CrossRef]
49. Greenewald, K.; Hero, A.O. Regularized block Toeplitz covariance matrix estimation via Kronecker product expansions. In Proceedings of the 2014 IEEE Workshop on Statistical Signal Processing (SSP), Gold Coast, Australia, 29 June–2 July 2014; pp. 9–12. [CrossRef]
50. Breloy, A.; Sun, Y.; Babu, P.; Ginolhac, G.; Palomar, D. Robust rank constrained kronecker covariance matrix estimation. In Proceedings of the 2016 50th Asilomar Conference on Signals, Systems and Computers, Pacific Grove, CA, USA, 6–9 November 2016; pp. 810–814. [CrossRef]
51. Xie, L.; He, Z.; Tong, J.; Liu, T.; Li, J.; Xi, J. Regularized Estimation of Kronecker-Structured Covariance Matrix. *arXiv* **2021**, arXiv:2103.09628.
52. Chen, Y.; Wiesel, A.; Hero, A. Robust Shrinkage Estimation of High-Dimensional Covariance Matrices. *IEEE Trans. Signal Process.* **2010**, *59*, 4097–4107. [CrossRef]
53. Pedregosa, F.; Varoquaux, G.; Gramfort, A.; Michel, V.; Thirion, B.; Grisel, O.; Blondel, M.; Prettenhofer, P.; Weiss, R.; Dubourg, V.; et al. Scikit-learn: Machine learning in Python. *J. Mach. Learn. Res.* **2011**, *12*, 2825–2830.
54. Virtanen, P.; Gommers, R.; Oliphant, T.E.; Haberland, M.; Reddy, T.; Cournapeau, D.; Burovski, E.; Peterson, P.; Weckesser, W.; Bright, J.; et al. SciPy 1.0: Fundamental Algorithms for Scientific Computing in Python. *Nat. Methods* **2020**, *17*, 261–272. [CrossRef] [PubMed]

55. Gramfort, A.; Luessi, M.; Larson, E.; Engemann, D.A.; Strohmeier, D.; Brodbeck, C.; Goj, R.; Jas, M.; Brooks, T.; Parkkonen, L.; et al. MEG and EEG Data Analysis with MNE-Python. *Front. Neurosci.* **2013**, *7*, 267. [CrossRef] [PubMed]
56. Pernet, C.R.; Appelhoff, S.; Gorgolewski, K.J.; Flandin, G.; Phillips, C.; Delorme, A.; Oostenveld, R. EEG-BIDS, an extension to the brain imaging data structure for electroencephalography. *Sci. Data* **2019**, *6*, 103. [CrossRef] [PubMed]
57. Appelhoff, S.; Sanderson, M.; Brooks, T.L.; Vliet, M.V.; Quentin, R.; Holdgraf, C.; Chaumon, M.; Mikulan, E.; Tavabi, K.; Höchenberger, R.; et al. MNE-BIDS: Organizing electrophysiological data into the BIDS format and facilitating their analysis. *J. Open Source Softw.* **2019**, *4*, 1896. [CrossRef]
58. Barachant, A. *MEG Decoding Using Riemannian Geometry and Unsupervised Classification*; Grenoble University: Grenoble, France, 2014.
59. Castaneda, M.H.; Nossek, J.A. Estimation of rank deficient covariance matrices with Kronecker structure. In Proceedings of the 2014 IEEE International Conference on Acoustics, Speech and Signal Processing (ICASSP), Florence, Italy, 4–9 May 2014; pp. 394–398. [CrossRef]
60. Blankertz, B.; Lemm, S.; Treder, M.; Haufe, S.; Müller, K.R. Single-trial analysis and classification of ERP components—A tutorial. *NeuroImage* **2011**, *56*, 814–825. [CrossRef]
61. Raudys, S.; Duin, R.P.W. Expected classification error of the Fisher linear classifier with pseudo-inverse covariance matrix. *Pattern Recognit. Lett.* **1998**, *19*, 385–392. [CrossRef]
62. Schäfer, J.; Strimmer, K. An empirical Bayes approach to inferring large-scale gene association networks. *Bioinformatics* **2004**, *21*, 754–764. [CrossRef]
63. Kraemer, N. On the Peaking Phenomenon of the Lasso in Model Selection. *arXiv* **2009**, arXiv:0904.4416.
64. Haufe, S.; Meinecke, F.; Görgen, K.; Dähne, S.; Haynes, J.D.; Blankertz, B.; Bießmann, F. On the interpretation of weight vectors of linear models in multivariate neuroimaging. *NeuroImage* **2014**, *87*, 96–110. [CrossRef]
65. Treder, M.S.; Blankertz, B. (C)overt attention and visual speller design in an ERP-based brain-computer interface. *Behav. Brain Funct.* **2010**, *6*, 28. [CrossRef] [PubMed]

Article

A Semi-Automatic Wheelchair with Navigation Based on Virtual-Real 2D Grid Maps and EEG Signals

Ba-Viet Ngo and Thanh-Hai Nguyen *

Faculty of Electrical and Electronics Engineering, Ho Chi Minh City University of Technology and Education, Ho Chi Minh City 700000, Vietnam
* Correspondence: nthai@hcmute.edu.vn

Abstract: A semi-automatic wheelchair allows disabled people to possibly control in an indoor environment with obstacles and targets. The paper proposes an EEG-based control system for the wheelchair based on a grid map designed to allow disabled people to reach any preset destination. In particular, the grid map is constructed by dividing it into grid cells that may contain free spaces or obstacles. The map with the grid cells is simulated to find the optimal paths to the target positions using a Deep Q-Networks (DQNs) model with the Parametric Rectified Linear Unit (PReLU) activation function, in which a novel algorithm for finding the optimal path planning by converting wheelchair actions is applied using the output parameters of the DQNs. For the wheelchair movement in one real indoor environment corresponding to the virtual 2D grid map, the initial position of the wheelchair will be determined based on natural landmarks and a graphical user interface designed for on-screen display can support disabled people in selecting the desired destination from a list of predefined locations using Electroencephalogram (EEG) signals by blinking eyes. Therefore, one user can easily and safely control the wheelchair using an EEG system to reach the desired target when the wheelchair position and destination are determined in the indoor environment. As a result, a grid map was developed and experiments for the semi-automatic wheelchair control were performed in real indoor environments to illustrate the effectiveness of the proposed method. In addition, the system is a platform to develop different types of controls depending on the types of user disabilities and different environmental maps built.

Keywords: EEG; BCI; graphical user interface; wheelchair navigation; grid map; natural landmark; optimal paths; deep Q-networks

1. Introduction

In recent years, many methods have been introduced to develop types of smart wheelchairs for people with different disabilities such as assistive technology [1], user physical interface [2], or semi-control (sharing control between user and machine) [3]. One of the most important problems in a smart wheelchair is to provide independent mobility for the elderly or severely disabled people, who cannot control an electric wheelchair using a joystick. Therefore, restoring their activity skills can significantly improve their life quality. The development of a typical smart wheelchair highly depends on the ability and disability of the user. It means that a patient with impaired activity often lacks muscle control and then it is difficult to control the movement of the arms and legs in the worst case. To support the mobility of patients, signals for control can be generated from actions such as voice, thoughts, eyes, and tongue [4–6]. In order to obtain good signals, users must control their emotions well and also highly concentrate for accuracy. This is difficult for users with severe disability, although it may be a good option. For people with severe disability, the best solution could use multiple signals from sensors installed on the user's body parts and the surrounding environment and the signals are analyzed before giving the desired commands for wheelchair control [7]. Using this solution could improve the difficulty of people with severe disability in the wheelchair control compared to solutions using one input.

EEG signals related to the human brain, which have challenging problems, have attracted many researchers. In particular, recent research on cognitive and motor control to improve the Brain–Computer Interface (BCI) for enhancing the health of elderly people has been shown [8]. According to this study, the BCI system can be useful for elderly people in training their motor/cognitive abilities to prevent the effects of aging. Therefore, it can help them to more easily control household appliances and to communicate information in daily activities. In [9], the authors represented the physical principles of BCI and the fundamental new methods for acquiring and analyzing EEG signals for controls related to brain activities. In particular, the BCI system was classified into three main categories including active, reactive, and passive. Regarding an active BCI, the neuro interface user controls a complex external device such as a wheelchair through a series of functional components of the control system and sees the results of this control on a screen. Reactive BCI inherits many features of active BCI, with a significant change to implement a control system based on the classification of brain responses to stimuli such as visuals, sounds, and touch. Passive BCIs are designed to monitor current brain activity and thereby provide important information about the operator's mental state, user intent, and situational interpretation. The Brain-Controlled Wheelchair (BCW) is a typical BCI application, which can help people with a physical disability to communicate with the outside environment. In [10], the BCW was exploited from many aspects, including the type of EEG signal acquisition, the command set for the control system, and the control method. Moreover, the authors summarized the recent development of the BCW and it can be mainly expressed in three aspects: from the wet electrode to the dry electrode; from single-mode to multi-mode; from synchronous control to asynchronous control. Therefore, it indicates that new functions have been employed in the BCW to increase its stability and robustness.

Mapping and navigation for wheelchairs or self-propelled robots have attracted many researchers in recent decades. The wheelchairs or self-propelled robots need to be provided maps for movement in detail so that they can be located in moving spaces. Moreover, their current coordinates were used as a basis for collecting new information during the moving process [11]. Mapping algorithms were gradually developed as Simultaneous Localization and Mapping (SLAM) algorithms and were applied to draw 3D maps [12], in which the computational problem of constructing or updating a map of an unknown environment is represented, with simultaneously keeping track of an agent's location within it. An image processing method was employed to identify fixed artificial landmarks built in moving space [13]. Thus, these fixed artificial landmarks were applied for determining the current location of a wheelchair on a map built in advance during its movement. Alcantarilla et al. proposed a powerful and fast method of positioning a wheelchair based on computer vision, in which image features were extracted and then combined with map components to provide a current position of self-propelled robots [14,15]. In fact, mapping for mobile robots in the environment is a major challenge due to the data obtained from the environment and the algorithm applied on them [16–18]. Landmark information for mobile robots plays an important role, in which types of landmarks [19–22] such as doors, stairs, walls, ceilings, and floors were selected and features were extracted for identification. Therefore, to detect the landmarks with their features, one could be based on color, texture, brightness, and obstacle size.

In recent years, the Reinforcement Learning (RL) method has achieved great success in many tasks including games [23] and simulation control agents [24]. Applications of the RL method in robot manufacturing are mostly limited in operation methods [25], in which the workspace could be fully observed and very stable. With mobile robots, complex environments can expand the sample space, while the RL method often takes action samples from a separate space for simpler processing of problems [26]. In [27], the RL method was applied for autonomous navigation based on the input of image information and has achieved significant success. The authors of this research analyzed agent behavior in static mazes with complex geometries, starting positions, and random orientations, while target positions could vary. The results showed that this RL method could allow the agent to navigate in large and intuitive environments, in which there were starting and target positions which were changed frequently, but the maze layout was always static. In [28], Yuke Zhu et al. tried to find the sequence of actions with minimum distance to move an agent from its current position to the target specified by the RGB image. This means that they have to collect a large number of different images to process before training the navigation model.

Finding a path on a static grid map is a well-known issue and well researched in AI communities, in which planners and robots with lots of methods and algorithms have been proposed to date [29,30]. Most of these algorithms are based on heuristic searching in the state space created by grid cells. In general, one prominent issue is that Neural Networks (NNs) work well with all types of tasks with data collected from sensors or images and then they are used as an input of the NN. The cells contain only two types of cells, including movable and non-movable, and they look like a perfect input for Modern Artificial Neural Network (MANN) architectures, such as a Convolutional Neural Network (CNN) and Recurrent Neural Network (RNN) [31–34]. In [23], the DQN algorithm was applied due to solving many challenges related to autonomous control [35]. In particular, this algorithm combines the Q-Learning algorithm with neural networks, in which DQNs solve problems with high-dimensional observation spaces using neural networks to estimate Q-values for corresponding actions. Mostly, deep RL training including DQNs and their variants is performed in a virtual environment, because the process of training using a trial and error method can lead to damage to real robots in typical tasks. The big difference between the structural simulation environment and the very complex real-world environment is the main challenge to directly transfer a trained model into a real robot.

In this article, an RL method applied to obtain results of the optimal path planning in a virtual 2D grid map is presented. In particular, in the first stage, the virtual 2D grid map is built based on a real environment, including free spaces, obstacles, landmarks, and targets. This virtual 2D grid map will be connected to the input of a DQN and the DQN's output is the Q-value of four actions (Right, Left, Up, Down) and the action with the largest Q-value will be selected so that the wheelchair can reach the desired target from any start point in the real environment. Therefore, in the second one, when the wheelchair moves in this real environment, it can use the scenery fully simulated as a Motion Planner (MP) through the virtual 2D grid map. Moreover, the wheelchair needs to determine its current position in both real and virtual environments with natural landmarks for movement. With the start and target positions determined, the MP will suggest the optimal path with control commands and a Wheelchair's Action Converter (WAC) will convert these control commands into actual control commands so that the wheelchair can complete its schedule. Finally, we evaluate the performance of the proposed model by performing a series of tests in simulation and in real environments. The results showed that the RL network architecture applied in this research to path-finding tasks is a potential issue in mobile vehicles in real environments based on landmarks, obstacles, and start and target points.

This article consists of four sections: Section 2 presents the structure of the system, the method for selecting destinations using EEG signals, and applying the RL algorithm for determining the optimal path of the wheelchair to the selected destination. In Section 3, the description of the basic specifications applied for wheelchair movement is given and the experiments related to the basic functions of the system and the experimental results using the proposed method are discussed. Finally, Section 4 presents the conclusions about this research.

2. Materials and Methods

2.1. System Architecture

In this research, one system architecture for an optimal path planning is proposed for wheelchair navigation to reach the desired targets. This system architecture includes two stages for the electric wheelchair in an indoor environment as described in Figure 1. In the first stage, the 2D grid maps with cells simulated based on one real indoor environment with different targets will provide information of cell states and targets' coordinates which are the inputs of DQNs. After being trained, the DQN model will have optimal parameters that can estimate the Q-values of all possible actions for that state. Therefore, the DQNs will have 4 outputs corresponding to 4 actions (Up, Down, Left, Right). Therefore, each 2D grid map is just built for one of the targets in one real indoor environment, so each DQN model is obtained for one MP.

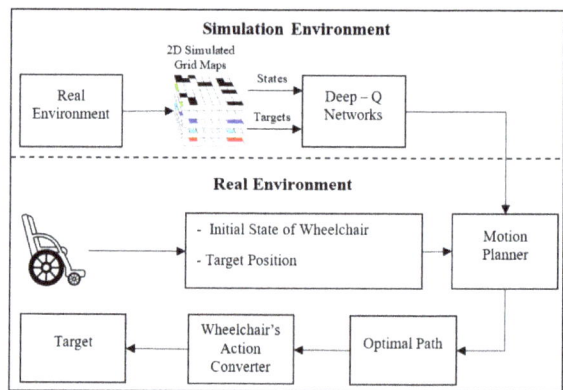

Figure 1. Representation of the system architecture for finding the optimal path of the wheelchair based on the 2D grid map.

The second stage is that the wheelchair will be controlled to reach the desired target in one real indoor environment. At the start time, the wheelchair will determine its start state itself based on natural landmarks and the desired target position in the real environment is known. When receiving the initial state of the wheelchair on the grid map, the DQN model will estimate the Q-values of 4 outputs corresponding to 4 actions (Up, Down, Left, Right). Therefore, the action with the highest Q-value will be selected. With this action, a new state on the grid map will be updated and then this new state will be the input to the DQN model and it will also select a corresponding action. This process will repeat and end when the state is the target. After navigating the optimal path to be able to reach the desired target, the MP with a sequence of actions (Right, Left, Up, Down) and the WAC will allow the wheelchair to move following this optimal path to reach that desired target.

In addition, as shown in Figure 1, the user needs to select a destination on the grid map using EEG signals. In the semi-automatic wheelchair system, the construction and the selection of destinations in a grid map for severely disabled people are a very important task. For people with severe disabilities not able to use normal controls, such as pressing a button, controlling a joystick, or touching a control screen, the EEG signal for controlling the semi-automatic wheelchair is a useful option. Using the EEG signal for directly controlling the semi-automatic wheelchair may cause stress due to concentrating for a long time, so the user can choose the desired destination through a screen interface with commands suitably designed for his/her actual environment [36]. The commands on the interface screen are assigned based on the type of the EEG signal from the user's face behaviors. Figure 2 describes the process of collecting, processing, and analyzing EEG signals for performing control commands related to the user interface. EEG signals are collected from an Emotiv EPOC system with 14 channels (14 electrodes) [37]. In particular, the EEG signals are collected from the electrodes located in the prefrontal cortex considered to be the most reliable signals. Therefore, the EEG signals are transferred to the signal pre-processing block for filtering and scaling before being sent to the feature extraction block. For the control of the wheelchair, the EEG signals after pre-processing are sent to the classification block for classifying input signals to produce control commands [36–38]. It means that the user can use the control commands for selecting one of destinations on the environmental map to reach.

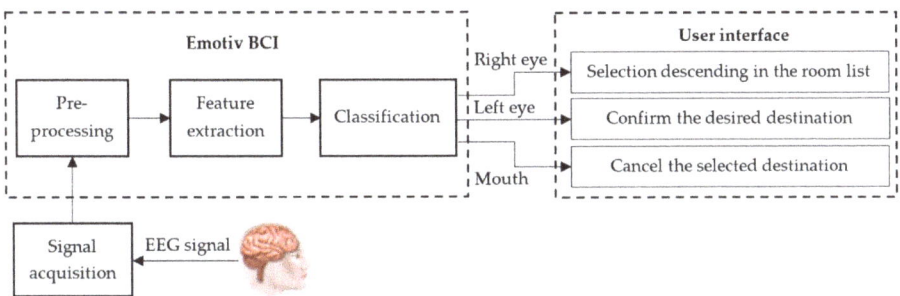

Figure 2. Brain–computer interface process flow.

The user interface is always designed to be simple and easy for disabled people, particularly, all commands can be operated using the BCI only as described in Figure 3. On the interface, the user will see a vertical menu with the symbols of destination names. The names in this menu are the pre-defined destinations such as living room, kitchen, and bedroom. To control the commands to reach the destinations, the act of closing the right eye of user is the command for selecting the desired destination. In particular, the user needs to close the right eye for 2 s to be able to move the cursor on the screen to the desired destination and then close the left eye to confirm the desired destination as shown in Figure 4. If the user wants to cancel the selected commands or cancel the selected destination, the user needs to perform the distortion of the mouth to the right. All operations selected for controlling the user interface with the designed destinations were tested on many users and the real results using the designed EEG commands produced the highest accuracy.

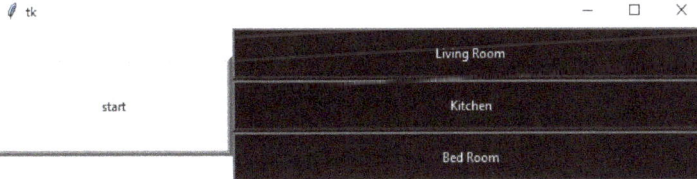

Figure 3. User interface for selecting the desired destination.

Figure 4. User interface selected the desired destination "Bed Room" using the EEG command.

2.2. Deep Q-Networks for Optimal Path Planning

In this study, DQNs were applied to find optimal paths as Q-tables based on virtual 2D grid maps through simulation, in which each target uses a virtual grid map and also many optimal paths are found for the wheelchair to reach that target from any start position of the wheelchair. In the DQNs, we could set variables related to the operation of the wheelchair and one real environment, particularly, the wheelchair is called Agent on a virtual 2D grid map (Environment), consisting of obstacles and free spaces. With the positions of start and target, Agent's task is to reach the target cell. In addition, the Agent interacts with Environment based on Actions (Left, Right, Up, Down). After each Action, Environment returns to Agent and State $S_t = (x_t, y_t)$ is the wheelchair position at time t, with the (x, y) grid coordinate, and the reward points (Reward, R) correspond to that State. In addition, Agent has a limited State, $S_t \in S$, with an $m \times n$ pre-defined size of S, and Agent is often placed in the middle of the grid cells for the possibility of moving in all four directions.

In this algorithm, State consists of three types of obstacle S_o, free space S_f, and target S_g. At each moment t, Agent is the State S_t and needs to select an Action from a fixed set of possible Actions. Therefore, the decision to select which Action for movement operation is only dependent on the current State, not the Action history, due to being irrelevant. In addition, the result of Action at time a_t will cause the conversion from the current State S_t at time t to the new State S_{t+1} at the time $(t + 1)$ and then immediate Reward collected after each Action $R(s_t, a_t) \in [-1, 1]$ is calculated using the following rule:

$$R(s_t, a_t) = \begin{cases} R_f & if\ a_t = s_t \to s_f \\ R_g & if\ a_t = s_t \to s_g \\ R_o & if\ a_t = s_t \to s_o \end{cases}. \quad (1)$$

Each movement of the wheelchair from one cell to an adjacent cell will lose R_f points and this will prevent it from wandering around and possibly reaching the desired target with the shortest path. In this algorithm, the maximum Reward is R_g points for movement of the wheelchair to hit the target. While the wheelchair tries to enter an obstacle cell, R_o points will be subtracted. It means that this is a serious punishment (penalty), so the wheelchair will learn how to completely avoid the punishment and so the effort to move to an obstacle cell is invalid and cannot be performed. The same rule for an attempt to move outside the map boundary with a punishment of R_b points applies. The case is that the wheelchair will lose R_p points for any movement to the cell that has been passed. Moreover, to avoid infinite loops during the training process using the DQNs, the total Reward is bigger than the negative threshold ($thr \times m \times n$) and then the wheelchair can move normally. Inversely, the movement of the wheelchair can be lost and many errors can be made, so the training needs to be carried out again until the total Reward is enough.

In this DQN, the main learning model is a Feedforward Neural Network (FWNN) with backpropagation training algorithm, in which the environmental States are the input of the network and bring Rewards back for each Action vector. The goal of Agent is to move following the map by a Policy to obtain a maximum Reward from the Environment. Therefore, Policy π at State s_t produces an Action a_t so that the total Reward Q Agent receives is the largest and is calculated by the following equation:

$$\pi(s_t) = \arg\max_{i=0,1,\dots,n} Q(s_t, a_i), \quad (2)$$

$$Q(s_t, a_t) = R(s_t, a_t) + \gamma.\max_{i=0,\dots,n} Q(s_{t+1}, a_i), \quad (3)$$

in which $Q(s_t, a_i)$ are Actions, a_i ($i = 0, 1, \dots, (n-1)$), n denotes the number of Actions and satisfies the following equation of Bellman [35], s_{t+1} is the next State, γ denotes the discount coefficient which makes sure that Agent is far from the target and it is smaller than the Q-value.

For approximating $Q(s_t, a_t)$, the FWNN has the input as a State and its output is the vector Q, in which the Q-value corresponds to n Actions. In addition, Q_i approximates the value of $Q(s_t, a_{ti})$ for each Action a_{ti}. When the network is fully and accurately trained, it will be used in the optimal path planning model for selecting Policy π as follows:

$$\pi(s_t) = a_j, \tag{4}$$

$$j = \arg\max_{i=0,\ldots,n}(Q_i), \tag{5}$$

in which the value j is determined based on the maximum Q.

The purpose of the neural network model is to learn how to exactly estimate the Q-value for Actions, so the objective/goal function applied here is to calculate the error Loss between the actual and predicted values Q and it is described by the following equation:

$$Loss = \left(R(s_t, a_t) + \gamma \max_{a_{t+1}} Q(s_{t+1}, a_{t+1}) - Q(s_t, a_t) \right)^2. \tag{6}$$

In addition, the FWNN model has the input of the current State and the outputs of the values Q. However, if the input of the FWNN is continually pushed into each State, it is very easily overfitted because the States are often the same and linear. For eliminating the overfitting problem in the FWNN model, a technique, called experience replay [23], is applied. In particular, instead of each State, the network is updated once, and the State is saved into memory and then sampled as small batches connected to the input of the FWNN for training. Therefore, it may provide diversification of the FWNN input and also avoid the overfitting problem. In this case, the training model will forget old samples not good enough for the training process and then they will be deleted from memory.

The FWNN model used in the training system has two hidden layers with the number of nodes equal to that of cells in the virtual 2D grid map built in the indoor environment. In addition, the size of the input layer is similar to the hidden one due to States of the virtual map used as the input. The output layer has the number of neurons equal to Actions (four Actions used in this paper) due to predicting the Q-value to estimate each Action. Finally, the FWNN model will choose the largest Q-value to perform an Action for the next State. In this research, the Parametric Rectified Linear Unit (PReLU) activation function, the optimization method of RMSProp, and the loss function of Mean Squared Error (MSE) are applied in the model for optimal path planning.

$$f(y_i) = f(y_i) = max(0, y_i) + a_i min(0, y_i) \tag{7}$$

in which y_i is any input on the ith layer and a_i is the negative slope which is a learnable parameter.

2.3. Wheelchair Navigation in Real Environment

In the optimal path planning, a simulated 2D grid map plays one very important role due to showing optimal paths for navigating the electric wheelchair to targets. In particular, one 2D grid map is simulated based on a lot of information related to one real environment. It means that the wheelchair, when moving in one real environment, may use parameters and State values for wheelchair navigation. Therefore, the simulated 2D grid map is divided into a lot of cells, including free spaces and occupancies. Each cell is calculated to be an actual area in the real environment with free spaces and obstacles and it can be one free space or one occupancy (obstacle). Therefore, we assume that the wheelchair can be driven through these free space areas to reach the desired target.

Figure 5 describes the 2D grid maps with occupancies and cells, including $m \times n$ cells in the indoor environment, in which the wheelchair can move through to reach targets. In particular, the real environment with objects (blue) is measured and divided into cells with the size of the wheelchair for creating the map as shown in Figure 5a. Therefore, the map with the divided cells is converted into a 2D grid map with calculation for filling cells related to obstacles (yellow). Therefore, the 2D grid map in Figure 5b is simulated to create the virtual 2D grid map as described in Figure 5c. The cells in the virtual 2D grid map are assigned 1 s to represent the occupied workspace (obstacles) and 0 s for the free workspace. Therefore, this virtual 2D grid map is considered as a binary map with black and white cells and the original coordinate of the virtual map is in the top left corner with the first location (0,0). It is obvious that this virtual map lets us know all cell locations which are used to find optimal paths using the algorithm of DQNs.

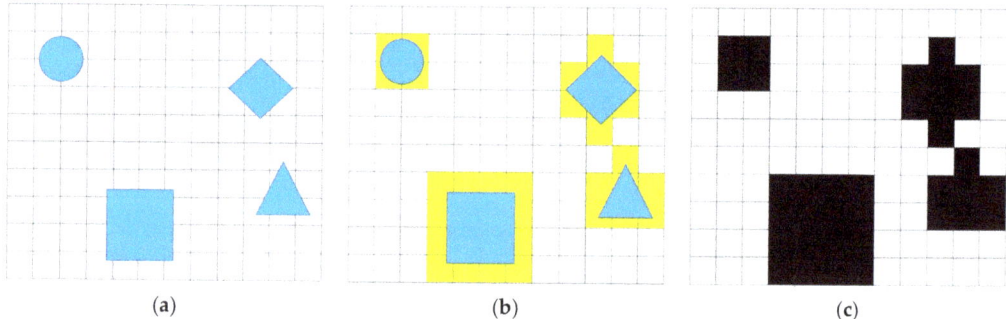

Figure 5. The occupancy 2D grid map of the real environment. (**a**) Environmental grid map with real obstacles and cells; (**b**) occupied cells related to the real obstacles; (**c**) virtual 2D grid map with black occupancy cells.

In this model, the wheelchair is located on a map through landmarks including the location and direction of the wheelchair on the map. The update of the position of the wheelchair is carried out when starting the movement for the first time. In this method, only the wheelchair location is connected to the input of the MP block for determining the optimal path and then it shows specific Actions with that State of the wheelchair.

One of the most important parts in the wheelchair control system is the wheelchair location in a real indoor environment for navigation. In a real indoor environment, natural landmarks will be automatically collected for creating one database for locating the motion wheelchair. In particular, the Features from Accelerated Segment Test (FAST) method is used to extract features of images captured from the camera system. Therefore, objects in the image that have the largest density of feature points are chosen to be natural landmarks and then the Speeded-Up Robust Features (SURF) algorithm is applied to identify these landmarks [39]. In this research, when the wheelchair is in the real environment as described in Figure 6, its initial location is determined based on three landmarks captured from a camera system installed on the wheelchair. Assume that the wheelchair moves in the flat space OXY with the unknown coordinates $W(x, y)$ and landmarks related to the coordinates in the real indoor environment. Therefore, obstacles selected as landmarks have distinctive characteristics which are different from other landmarks with their coordinates $A(x_A, y_A)$, $B(x_B, y_B)$, and $C(x_C, y_C)$ [40]. The wheelchair position can be determined if the coordinates of the landmarks and the corresponding distances from the wheelchair to the landmarks are known. Based on the wheelchair location determined as above, the wheelchair position on the real grid map with the square cell size $(a \times a)$ is $S_w(\frac{x}{a}, \frac{y}{a})$.

The starting point $S_W(1,0) \in S_f$ and the target $T_i(3,2) \in T$ are obtained based on the pre-trained map with this target, in which S_f is a set of free cells and T is a set of known targets. Therefore, the MP gives one optimal path which is a set of Actions including Right, Right, Down, Down as shown in Figure 6. It means that the wheelchair impossibly moves based on these Actions due to the wheelchair model in this research not being an omnidirectional control model. In Figure 7a, the two-input converting block is Action a, determined from the MP output, and the initial direction d of the wheelchair includes the four directions (Up, Down, Left, Right) as described in Figure 7b. Thus, the output of the converting block a_w is an Action that is suitable with the wheelchair orientation/direction in the real environment.

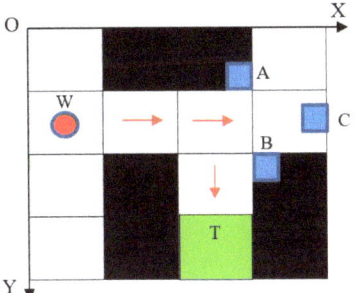

Figure 6. Coordinates of the wheelchair, landmarks, and target in simulated 2D grid map.

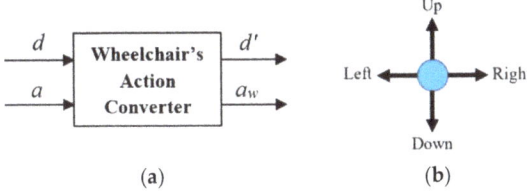

Figure 7. The representation of converting actual control commands from the simulation. (**a**) Converter with the simulated inputs and the actual outputs; (**b**) representation of four control directions.

The training process for finding the optimal path will produce a series of Actions with different States, in which these Actions will produce many optimal paths dependent on the initial position of the wheelchair. Therefore, after each Action a, the wheelchair direction d will change into a new direction d'. For the movement of the wheelchair, we propose a novel algorithm based on the WAC as described in Figure 7. In particular, the wheelchair Actions a_w and the new direction $d' = a$ during its movement in real environment need to be determined and this algorithm is expressed as follows:

$$\begin{cases} a_w = \begin{cases} Forward & if \quad a = Up \\ Backward & if \quad a = Down \\ Left - Forward & if \quad a = Left \\ Right - Forward & if \quad a = Right \end{cases} \\ d = Up \end{cases}, \qquad (8a)$$

$$\begin{cases} a_w = \begin{cases} Forward & if \quad a = Down \\ Backward & if \quad a = Up \\ Left - Forward & if \quad a = Right \\ Right\, and\, Forward & if \quad a = Left \end{cases} \\ d = Down \end{cases}, \qquad (8b)$$

$$\begin{cases} a_w = \begin{cases} Forward & if \quad a = Left \\ Backward & if \quad a = Right \\ Left - Forward & if \quad a = Down \\ Right - Forward & if \quad a = Up \end{cases} \\ d = Left \end{cases}, \qquad (8c)$$

$$\begin{cases} a_w = \begin{cases} Forward & if \ a = Right \\ Backward & if \ a = Left \\ Left - Forward & if \ a = Up \\ Right - Forward & if \ a = Down \end{cases} \\ d = Right \end{cases} \quad , \tag{8d}$$

in which a and d are the parameters which are determined based on Action and direction in MP. In Equations (8a)–(8d), the wheelchair Actions a_w are defined as follows:

- a_w = Forward: The wheelchair will go straight;
- a_w = Backward: The wheelchair will go back;
- a_w = Left-Forward: The wheelchair will rotate left and then go straight;
- a_w = Right-Forward: The wheelchair will rotate right and then go straight;
- a_w = Stop if a = no Action: The wheelchair will stop.

3. Results and Discussion

3.1. Simulation of Path Training for the Wheelchair Based on 2D Grid Map

We constructed two grid maps depicting the indoor environment as shown in Figure 8, where the white cells are the spaces, the black cells are the obstacles, and the red cells are the targets. During training and testing the proposed structure, the PC configuration with the Windows operating system was Intel (R) Core (TM) i5-6300U, 2.4 GHz, 16 GB RAM. During each training, the starting position is randomly selected in the map and guaranteed not to overlap with the obstacle cell. Table 1 describes the parameters which are trained in the case as described in Figure 8.

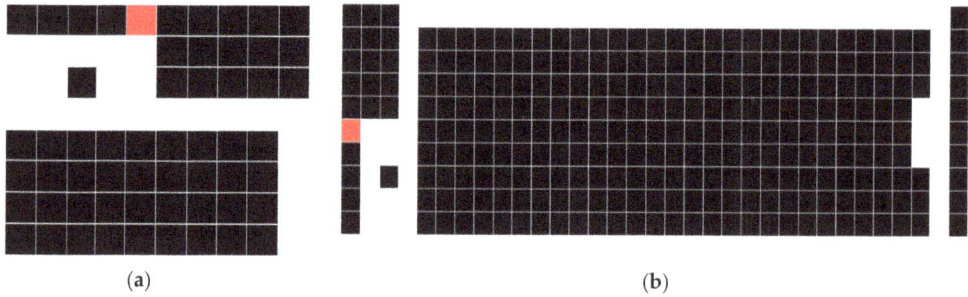

Figure 8. Training environment simulated using the proposed model. (**a**) An 8 × 11 grid map; (**b**) 11 × 33 grid map.

Table 1. Training Parameters.

Parameter	Value
Learning rate	0.00001
Discount factor γ	0.8
Exploration	0.1
Mini-batch size	32
Replay memory size	100
Reward when moving outside the map R_b	−0.8
Reward of free space R_f	−0.4
Reward of obstacle R_o	−0.75
Reward of goal R_g	1

To evaluate the effectiveness of the DQN method, we performed the experiment with different steps and different environments, and the stable results of the DQN method are shown in Figures 9 and 10 for each environment. In particular, we worked out the experiment of the proposed model using DQNs with two activations of PReLU and ReLU for comparing the performance between them, where the horizontal axis is the number of episodes and the vertical axis is the Win rate. The Win rate is calculated based on the number of Wins per the total number of selected positions to start a game in an episode. From Figure 9, we can see that the Win rate can increase or decrease or stay the same after each episode.

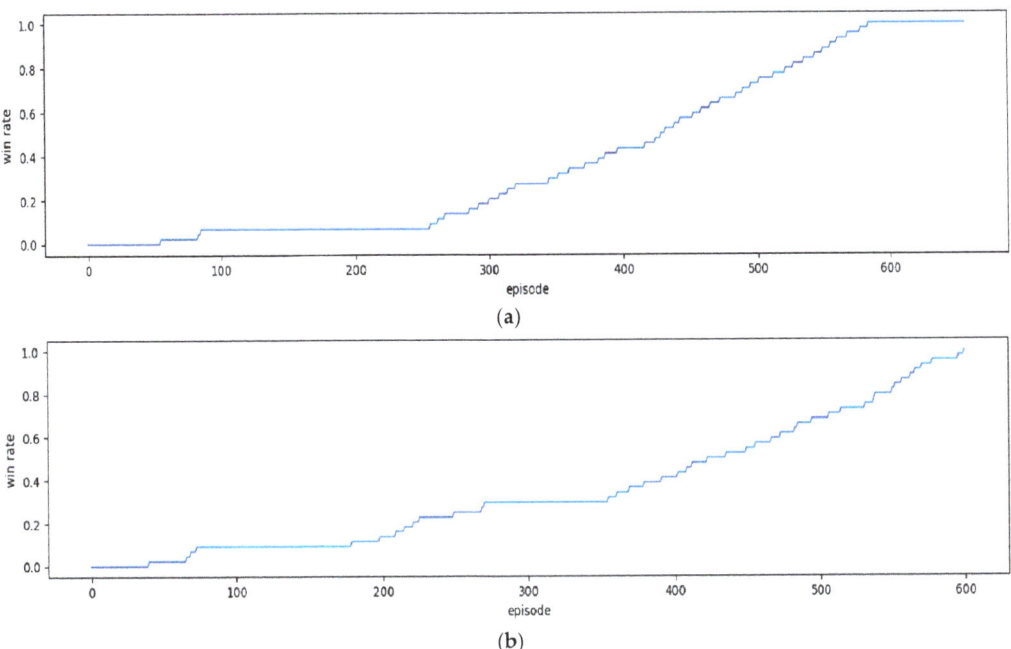

Figure 9. The comparison of Win rates when training the DQN model with two activation types in the case of the 8 × 11 grid map. (**a**) The DQN model with PReLU activation; (**b**) the DQN model with ReLU activation.

According to the results in Figure 9 with a small environment, the two models of DQNs-PReLU and DQNs-PReLU have the same Win rate growth path and also reach the maximum Win rate threshold of 1 after about 600 episodes. Figure 10 shows the Win rate growth of the large environment with the two selected models. With the results of the model of DQNs-PReLU in Figure 10a, when the episode is over 7000, the Win rate starts sharply increasing and then reaches the maximum threshold at episode 15,000. Therefore, the Win rate reaches saturation and this shows that the model meets the training requirements and then ends. In contrast, according to the results shown in Figure 10b using the model of DQNs-ReLU, the Win rate starts sharply increasing when the episode is over 25,000 and reaches the maximum threshold when the episode is 240,000. After that, the Win rate reaches saturation and this means that the model meets the requirements of training and then ends. Thus, it can be seen that in a large environment, the model of DQNs-PReLU more quickly reaches the maximum score than DQNs-ReLU.

In addition, the obtained results are comparable in terms of training time and the number of episodes of the DQN model with the two types of activations as shown in Table 2. In particular, in the small environment with 8 × 11, the difference in training time is not too large, 36.3 s compared to 42.3 s for two ReLU and PReLU activations, respectively. With the episode number of the two models of DQNs-ReLU and DQNs-PReLU used for training, this environment is not much different, with episode numbers of 601 and 607, respectively. However, with the larger environment of 11 × 33, there is a big difference in training time and the number of episodes between the two models. In particular, the training time of the DQNs-ReLU model is nearly 4 times larger than that of the DQNs-PReLU model. In addition, the average number of episodes per training time using the DQNs-ReLU model is 15 times that of the DQNs-PReLU. This means that the DQNs-PReLU model gives better performance than DQNs-ReLU using this environment.

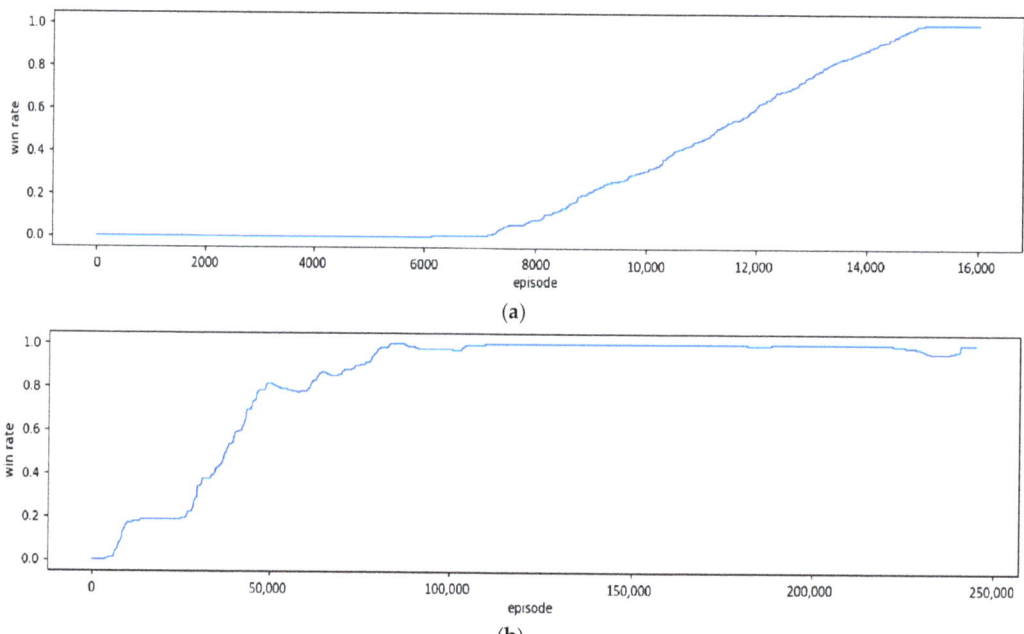

Figure 10. The comparison of Win rates when training the DQN model with two activation types in the case of the 11 × 33 grid map. (**a**) The DQN model with PReLU activation; (**b**) the DQN model with ReLU activation.

Table 2. The Relative Performance of Proposed DQN Models.

Environment	Model	Average No. of Episodes	Average Training Time
Small (8 × 11)	DQNs with ReLU activation	601.0	36.3 s
	DQNs with PReLU activation	657.0	42.3 s
Large (11 × 33)	DQNs with ReLU activation	244,879	6.05 h
	DQNs with PReLU activation	16,015	35.24 min

Table 3 describes the comparison of episode and time using the DQN model with two activations and previous models in training the two environments (small and large). In all experiments of randomly trained models, we performed training of each case 10 times to take the average training time and the average number of episodes. It is obvious that the Traditional Q-Learning model shows a table to record the value of each pair (State, Action), in which the State with the highest value indicates the most desirable Action. Therefore, these values are constantly refined during training and this is a quick way to learn a Policy. The second model, called the SARSA model, uses a setup similar to the previous model, but takes fewer risks during learning. During the training process, depending on the small or large environment, the training time and the number of episodes will be different.

Table 3. The Relative Performance of Previous Models.

Environment	Model	Average No. of Episodes	Average Training Time
Small (8 × 11)	Traditional Q-Learning [41]	60.0	198.4 s
	SARSA [42]	75.0	223.9 s
Large (11 × 33)	Traditional Q-Learning [41]	235.0	1.45 h
	SARSA [42]	275.0	57.23 min

In particular, with a small environment, the training time and the number of episodes are less than those with a large environment as shown in Tables 2 and 3. Furthermore, in Table 3, the models have a small number of episodes and a lot of time because Traditional Q-Learning works based on finding the maximum reward for each step and the larger the number of States, the larger the Q-table, so the calculation will take a lot of time. Meanwhile, in Table 2, the DQN has a lot of episodes but it takes less computation time because the DQN chooses some random and risky decisions to quickly obtain a high reward and it will accept to lose a certain amount of episodes.

With the statistical results in Tables 2 and 3, although the number of episodes in the training process is much larger than that of the Q-table-based models in Table 3, the DQNs-PReLU model in Table 2 takes a longer training time in two training cases for both small and large environments. In particular, for the small environment, the model of DQNs-PReLU has about 10 times more episodes than the models of Traditional Q-Learning and SARSA, but its training time is almost 5 times less than that of the Traditional Q-Learning and SARSA. In addition, with a large environment, DQNs-PReLU has a large number of about 16,015 episodes, nearly 60 times more than that using the Traditional Q-Learning, and nearly 70 times more than that using the SARSA model. However, the training time is significantly reduced with about 35.24 min compared to that of two models in Table 3, 1.45 h and 57.23 min, respectively. As an extra feature after learning, it saves the model to disk so this can be loaded later for the next game. Therefore, a neural network needs to be used in a real-world situation where training is separated from actual use.

3.2. Wheelchair Movement to Reach Map-Based Desired Target

The experiment was performed in an environment of 126.72 m^2 which was divided into square grids of one map with a size of 8 × 11, in which each square has a size of 1.2 m × 1.2 m as shown in Figure 11. The wheelchair was installed to be able to move at the speed of 3 km/h for matching the processing speed of the system. An electrical wheelchair was installed with an RGB-D camera system and other equipment as shown in Figure 12. Information about the surrounding environment obtained from the camera system was processed by a computer and then transferred to the motor system of the wheelchair for motion control. In addition, in this research, we performed two experiments, including a self-control user and an automatic control user. In the self-control user model, the user can self-control commands such as going forward, backward, and turning right and left during the wheelchair movement. Meanwhile, the automatic control user mode means that the user can choose one of the targets by using EEG signals which are assigned to the targets to reach [36] and our proposed algorithm in the wheelchair control system is applied so that the wheelchair can automatically reach the chosen target.

Figure 13 shows the green real path of the wheelchair, which was controlled by the user during reaching the target. In particular, the discontinuous green path is the desired path in the real environment that the wheelchair needs to follow to reach the target, while the red path of the wheelchair is the path controlled by self-control mode using EEG signals [38] to go straight, turn left and right during reaching the destination. With the experiment using the self-control, the wheelchair moved according to the red path and then turned to the undesired direction shown by the red path and blue dash-dot ellipse. It means that in this case, the wheelchair could very easily have an obstacle collision. In addition, with the mode of the self-control, the movement of the wheelchair is unstable and discontinuous as shown in Figure 13. In particular, the wheelchair went straight, then stopped, turned right, and then was continuously interrupted during the movement time. It is obvious that the user was trying hard to control it to turn right or left and go straight.

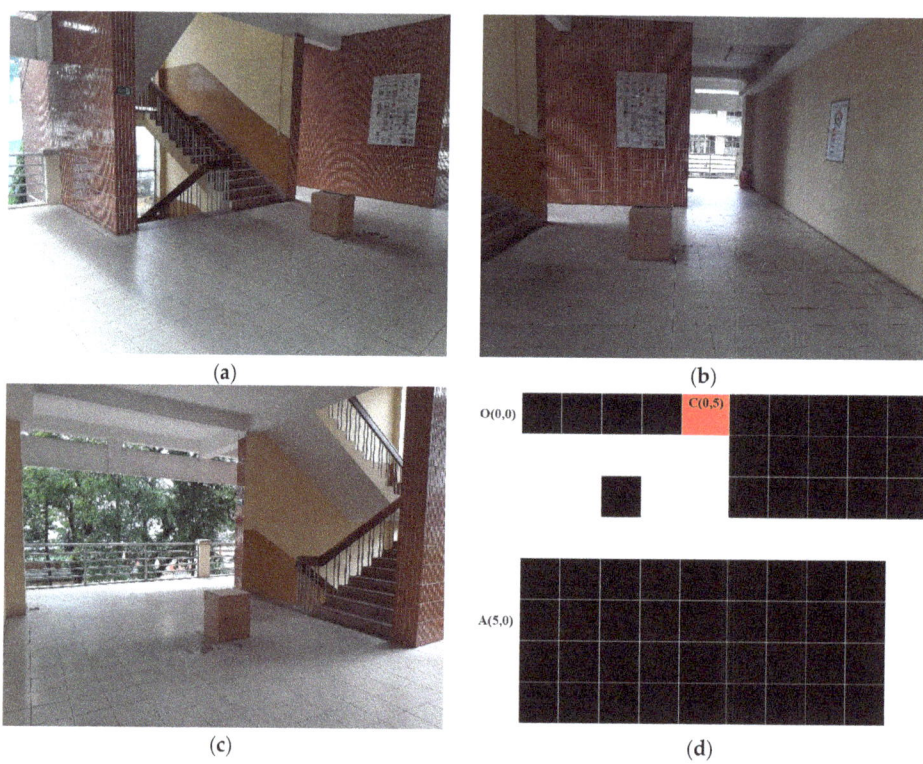

Figure 11. The experimental environment. (**a**) The 1st view of the real environment; (**b**) the 2nd view of the real environment; (**c**) the 3rd view of the real environment; (**d**) the 2D grid map.

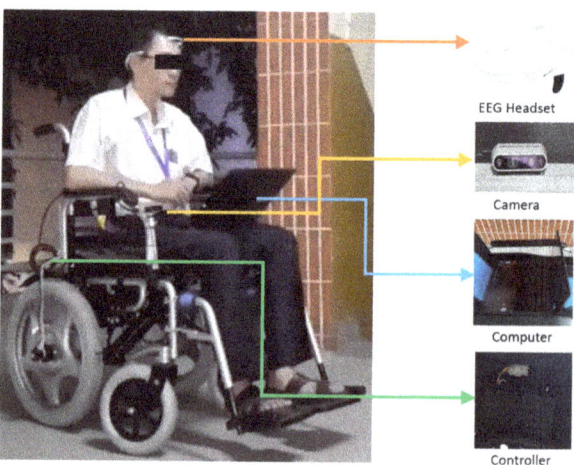

Figure 12. The wheelchair navigation system installed with devices.

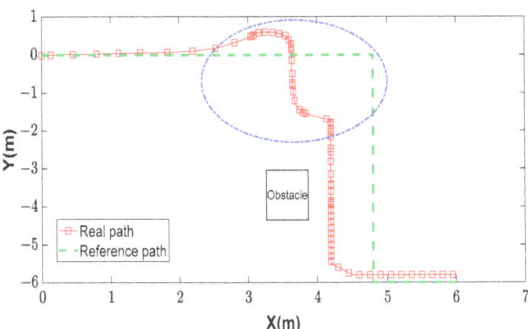

Figure 13. The real path of the wheelchair movement and the reference path.

For improving the wheelchair control using the self-control mode, we used the proposed model with the semi-automatic control. With this mode, the user just needs to choose one typical target by using EEG commands and then the wheelchair will automatically move to reach the desired target with high stability and smoothness. In particular, using the environmental map in Figure 11a–c, the actual paths of the wheelchair after moving to reach the target were as shown in Figure 14b. Therefore, the moving process was re-calculated and the path positions of the wheelchair with the axes of X and Y were re-drawn for the purpose of the comparison with the simulation paths (blue arrows) as shown in Figure 14a. The starting point of the wheelchair is random and the wheelchair automatically determines its position on the map by identifying landmarks in the environment. In particular, in this case, the wheelchair determined it position on the grid map at the coordinate A(5,0) and the direction of the wheelchair d is Up. In the semi-automatic wheelchair, people with disabilities can control the wheelchair using EEG signals to select one of the commands on the interface screen with one sign corresponding to the target C(0,5). With the starting point A(5,0) and the target point C(0.5) selected, the RL model will produce a sequence of control commands for the path and then these commands are converted to control commands in the real environment for the wheelchair using Equations (8a)–(8d) as shown in Table 4.

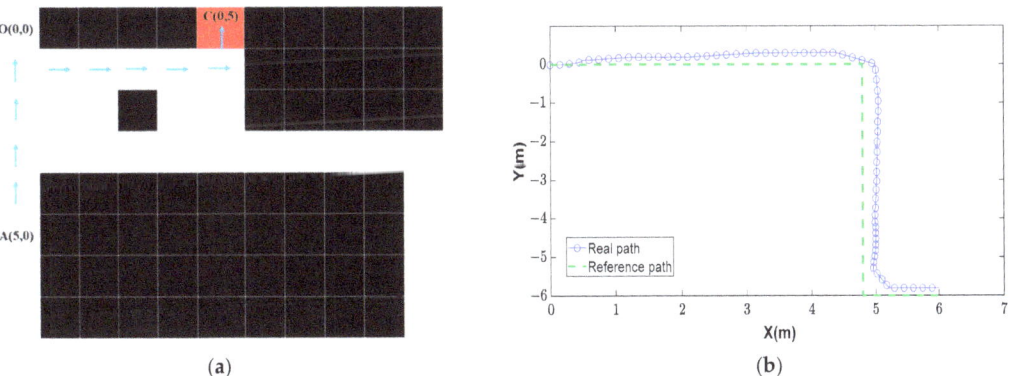

Figure 14. Representation of the simulation route using the semi-automatic control and the wheelchair's real path (**a**) The blue arrow route simulated using DQNs; (**b**) the wheelchair movement path in the real environment using DQNs and the reference path.

Table 4. Wheelchair Control Commands Converted from Simulation Commands.

State of Wheelchair	Current Direction D	Action of Model A	New Direction d'	Action of Wheelchair a_w
(5,0) to (4,0)	Up	Up	Up	Forward
(4,0) to (3,0)	Up	Up	Up	Forward
(3,0) to (2,0)	Up	Up	Up	Forward
(2,0) to (1,0)	Up	Up	Up	Forward
(1,0) to (1,1)	Up	Right	Right	Right–Forward
(1,1) to (1,2)	Right	Right	Right	Forward
(1,2) to (1,3)	Right	Right	Right	Forward
(1,3) to (1,4)	Right	Right	Right	Forward
(1,4) to (1,5)	Right	Right	Right	Forward
(1,5) to (0,5)	Right	Up	Up	Left–Forward

In addition, in this experiment, the actual path of the wheelchair with the proposed method of DQNs (blue path) is compared with the standard path (green dashed path), as shown in Figure 14b, for evaluating the wheelchair movement path and the simulated path. The results showed that the wheelchair could move to reach the desired target with the average error of ±0.2 m in the X axis and ±0.2 m in the Y axis.

The purpose of these experiments is to compare the results of the semi-automatic control using the RL method with the self-control by the user using the EEG signals. In particular, Figure 15a shows three graphs which represent the wheelchair movements, in which the blue path is that of the proposed mode and the red path is that of the self-control mode. From Figure 15a, it can be seen that the wheelchair's path when controlled by the semi-automatic control method is closer to the reference path than when using the self-control method. In addition, the wheelchair path using the semi-automatic control is smoother and more continuous than the path using the self-control. To clarify the two control methods, we recorded the wheelchair control commands during the movement to reach the destination.

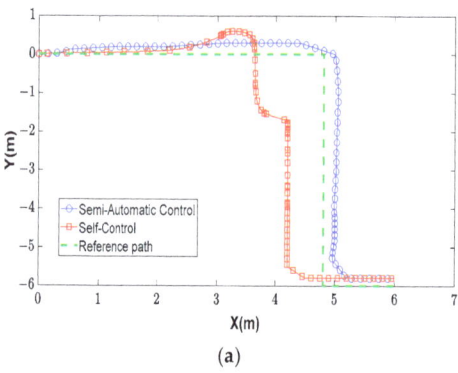
(a)

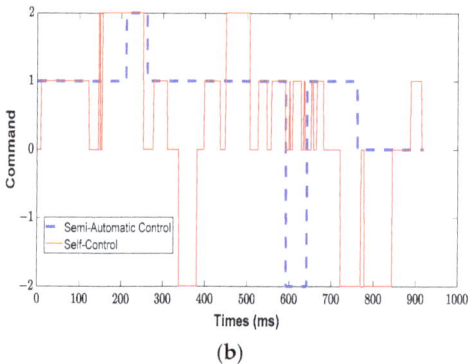
(b)

Figure 15. The comparison of the stable movements of the wheelchair in two control methods (semi-automatic control and self-control). (a) The real paths of the two control methods and the reference path; (b) the control sequences of the two control methods.

In Figure 15b, the control commands are shown on the vertical axis with the values of −2, 0, 1, 2 corresponding to the commands to turn left, stop, go straight, and turn right. Therefore, it could be seen that the wheelchair moved with high stability in the case of the semi-automatic control with different movement environments compared to the mode of the self-control user. In addition, the result showed that the automatic control user mode spends less time on wheelchair movement with the average of about 80 s compared to that of the self-control user with the average time of about 95 s.

In another case, Figure 16d shows the simulation paths (blue arrows) of the wheelchair based on the environmental map in Figure 16a–c when the wheelchair moves from O(0,0) to C(0,5). From Figure 16e, it can be seen that the wheelchair's path controlled by the semi-automatic control method is shorter and smoother compared to the self-control method. Further, the semi-automatic control method has an average error of 0.1 m in the X axis and 0.3 m in the Y axis compared with ± 0.5 m in the X axis and ± 0.5 m in the Y axis of the self-control method. With Figure 16f, it can be seen that the wheelchair moved with high stability in the case of the semi-automatic control with different movement environments compared to the mode of the self-control user.

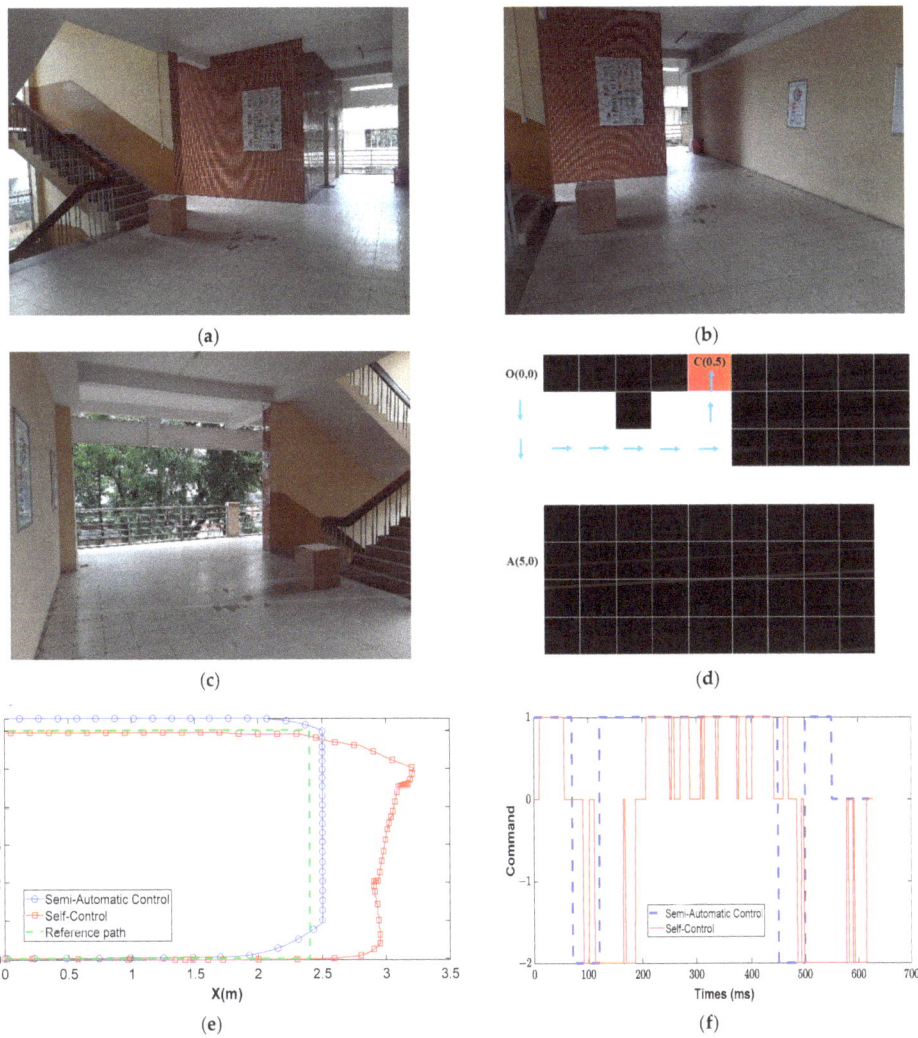

Figure 16. The comparison of the stable movements of the wheelchair in two control methods (semi-automatic control and self-control). (**a**) The 1st view of the real environment; (**b**) the 2nd view of the real environment; (**c**) the 3rd view of the real environment; (**d**) the blue arrow route simulated using DQNs; (**e**) the real paths of the two control methods and the reference path; (**f**) the control sequences of the two control methods.

4. Conclusions

The paper presents a semi-control method of an electric wheelchair combined with an RGB-D camera system, a graphical user interface, and real environmental maps with natural landmarks, in which optimal path planning for the wheelchair navigation was determined. In particular, 2D grid maps were used for training to create the shortest paths to the targets, in which the virtual-real RL method using DQNs carried out the training process effectively. After training, disabled people may select the desired target on the interface-user map using EEG signals to reach it. Therefore, the semi-control wheelchair located itself based on natural landmarks during movement following the optimal path from the motion planner in the real indoor environment. With the proposed method for the optimal path based on DQNs, the semi-control wheelchair could operate well to reach the desired target with small errors compared to the simulation trajectory, as well as to the trajectory of the self-control user using an EEG system. It is obvious that, with our proposed optimal path trajectory and the semi-automatic control method, the semi-control wheelchair movement is more stable, safe, and takes less time for moving. As a result of the proposed method, this wheelchair control system can be developed to apply to more complex environments with obstacles in the future.

Author Contributions: Conceptualization, methodology, simulation, and writing original draft, B.-V.N.; methodology, supervision, validation and writing—review and editing, T.-H.N. All authors have read and agreed to the published version of the manuscript.

Funding: This research received no external funding.

Institutional Review Board Statement: Not applicable.

Informed Consent Statement: Not applicable.

Data Availability Statement: Not applicable.

Acknowledgments: We would like to thank Ho Chi Minh City University of Technology and Education, Vietnam.

Conflicts of Interest: The authors declare no conflict of interest.

References

1. Kim, E.Y. Wheelchair Navigation System for Disabled and Elderly People. *Sensor* **2016**, *16*, 1806. [CrossRef] [PubMed]
2. Xu, X.; Zhang, Y.; Luo, Y.; Chen, D. Robust Bio-Signal Based Control of an Intelligent Wheelchair. *Sensors* **2013**, *2*, 187–197. [CrossRef]
3. Xi, L.; Shino, M. Shared Control of an Electric Wheelchair Considering Physical Functions and Driving Motivation. *Int. J. Environ. Res. Public Health* **2020**, *17*, 5502. [CrossRef]
4. Kim, J.; Park, H.; Bruce, J.; Sutton, E.; Rowles, D.; Pucci, D.; Holbrook, J.; Minocha, J.; Nardone, B.; West, D.; et al. The tongue enables computer and wheelchair control for people with spinal cord injury. *Sci. Trans. Med.* **2013**, *5*, 1–22. [CrossRef] [PubMed]
5. Dahmani, M.; Chowdhury, M.; Khandakar, A.; Rahman, T.; Al-Jayyousi, K.; Hefny, A.; Kiranyaz, S. An Intelligent and Low-Cost Eye-Tracking System for Motorized Wheelchair Control. *Sensors* **2020**, *20*, 3936. [CrossRef] [PubMed]
6. Shahin, M.K.; Tharwat, A.; Gaber, T.; Hassanien, A.E. A Wheelchair Control System Using Human-Machine Interaction: Single-Modal and Multimodal Approaches. *J. Intell. Syst.* **2019**, *28*, 115–132. [CrossRef]
7. Barriuso, A.L.; Pérez-Marcos, J.; Jiménez-Bravo, D.M.; Villarrubia González, G.; de Paz, J.F. Agent-Based Intelligent Interface for Wheelchair Movement Control. *Sensors* **2018**, *18*, 1511. [CrossRef]
8. Belkacem, A.N.; Jamil, N.; Palmer, J.A.; Ouhbi, S.; Chen, C. Brain Computer Interfaces for Improving the Quality of Life of Older Adults and Elderly Patients. *Front. Neurosci.* **2020**, *14*, 692. [CrossRef]
9. Hramov, A.E.; Maksimenko, V.A.; Pisarchik, A.N. Physical principles of brain–computer interfaces and their applications for rehabilitation, robotics and control of human brain states. *Phys. Rep.* **2021**, *918*, 1–133. [CrossRef]
10. Wang, H.; Yan, F.; Xu, T.; Yin, H.; Chen, P.; Yue, H.; Bezerianos, A. Brain-Controlled Wheelchair Review: From Wet Electrode to Dry Electrode, From Single Modal to Hybrid Modal, From Synchronous to Asynchronous. *IEEE Access* **2021**, *9*, 55920–55938. [CrossRef]
11. Fuentes-Pacheco, J.; Ruiz-Ascencio, J.; Rendón-Mancha, J.M. Visual simultaneous localization and mapping: A survey. *Artif. Intell. Rev.* **2015**, *43*, 55–81. [CrossRef]

12. Cras, J.L.; Paxman, J. A modular hybrid SLAM for the 3D mapping of large scale environments. In Proceedings of the 12th International Conference on Control Automation Robotics & Vision (ICARCV), Guangzhou, China, 5–7 December 2012; pp. 1036–1041.
13. Han, S.B.; Kim, J.H.; Myung, H. Landmark-Based Particle Localization Algorithm for Mobile Robots with a Fish-Eye Vision System. *IEEE/ASME Trans. Mechatron.* **2013**, *18*, 1745–1756. [CrossRef]
14. Alcantarilla, P.F.; Oh, S.M.; Mariottini, G.L.; Bergasa, L.M.; Dellaert, F. Learning visibility of landmarks for vision-based localization. In Proceedings of the IEEE International Conference on Robotics and Automation (ICRA), Anchorage, AK, USA, 3–7 May 2010; pp. 4881–4888.
15. Wang, H.; Ishimatsu, T. Vision-Based Navigation for an Electric Wheelchair Using Ceiling Light Landmark. *J. Intell. Robot. Syst.* **2004**, *41*, 283–314. [CrossRef]
16. Hu, G.; Huang, S.; Zhao, L.; Alempijevic, A.; Dissanayake, G. A robust RGB-D SLAM algorithm. In Proceedings of the IEEE/RSJ International Conference on Intelligent Robots and Systems, Vilamoura-Algarve, Portugal, 7–12 October 2012; pp. 1714–1719.
17. Basu, A.; Ghosh, S.K.; Sarkar, S. Autonomous navigation and 2D mapping using SONAR. In Proceedings of the 5th International Conference on Wireless Networks and Embedded Systems (WECON), Rajpura, India, 14–16 October 2016; pp. 1–5.
18. Liu, X.; Guo, B.; Meng, C. A method of simultaneous location and mapping based on RGB-D cameras. In Proceedings of the 14th International Conference on Control, Automation, Robotics and Vision (ICARCV), Phuket, Thailand, 13–15 November 2016; pp. 1–5.
19. Zhong, X.; Zhou, Y.; Liu, H. Design and recognition of artificial landmarks for reliable indoor self-localization of mobile robots. *Int. J. Adv. Robot. Syst.* **2017**, *14*, 1–13. [CrossRef]
20. Yu, T.; Shen, Y. Asymptotic Performance Analysis for Landmark Learning in Indoor Localization. *IEEE Commun. Lett.* **2018**, *22*, 740–743. [CrossRef]
21. Viet, N.B.; Hai, N.T.; Hung, N.V. Tracking landmarks for control of an electric wheelchair using a stereoscopic camera system. In Proceedings of the International Conference on Advanced Technologies for Communications (ATC 2013), Ho Chi Minh City, Vietnam, 16–18 October 2013; pp. 339–344.
22. Montero, A.S.; Sekkati, H.; Lang, J.; Laganière, R.; James, J. Framework for Natural Landmark-based Robot Localization. In Proceedings of the 9th Conference on Computer and Robot Vision, Toronto, ON, Canada, 28–30 May 2012; pp. 131–138.
23. Mnih, V.; Kavukcuoglu, K.; Silver, D.; Rusu, A.A.; Veness, J.; Bellemare, M.G.; Graves, A.; Riedmiller, M.; Fidjeland, A.K.; Ostrovski, G.; et al. Human-level control through deep reinforcement learning. *Nature* **2015**, *518*, 529–533. [CrossRef]
24. Lillicrap, T.P.; Hunt, J.J.; Pritzel, A.; Heess, N.; Erez, T.; Tassa, Y.; Silver, D.; Wierstra, D. Continuous control with deep reinforcement learning. In Proceedings of the ICLR 2016, San Juan, Puerto Rico, 2–4 May 2016.
25. Gu, S.; Lillicrap, T.; Sutskever, I.; Levine, S. Continuous deep Q-Learning with model-based acceleration. In Proceedings of the 33rd International Conference on Machine Learning, New York, NY, USA, 19–24 June 2016; pp. 2829–2838.
26. Khan, M.U. Mobile Robot Navigation Using Reinforcement Learning in Unknown Environments. *Balk. J. Electr. Comput. Eng.* **2019**, *7*, 235–244. [CrossRef]
27. Panov, A.I.; Yakovlev, K.S.; Suvorov, R. Grid Path Planning with Deep Reinforcement Learning: Preliminary Results. *Proc. Comput. Sci.* **2018**, *123*, 347–353. [CrossRef]
28. Zhu, Y.; Mottaghi, R.; Kolve, E.; Lim, J.J.; Gupta, A.; Fei-Fei, L.; Farhadi, A. Target-driven visual navigation in indoor scenes using deep reinforcement learning. In Proceedings of the 2017 IEEE International Conference on Robotics and Automation, Singapore, 29 May–3 June 2017; pp. 3357–3364.
29. Lei, X.; Zhang, Z.; Dong, P. Dynamic Path Planning of Unknown Environment Based on Deep Reinforcement Learning. *J. Robot.* **2018**, *2018*, 1–10. [CrossRef]
30. Konar, A.; Chakraborty, I.G.; Singh, S.J.; Jain, L.C.; Nagar, A.K. A Deterministic Improved Q-Learning for Path Planning of a Mobile Robot. *IEEE Trans. Syst. Man Cybern. Syst.* **2013**, *43*, 1141–1153. [CrossRef]
31. Quan, H.; Li, Y.; Zhang, Y. A Novel Mobile Robot Navigation Method Based on Deep Reinforcement Learning. *Int. J. Adv. Robot. Syst.* **2020**, *17*, 1–11. [CrossRef]
32. Duryea, E.; Ganger, M.; Hu, W. Exploring Deep Reinforcement Learning with Multi Q-Learning. *Intell. Control Autom.* **2016**, *7*, 129–144. [CrossRef]
33. Bae, H.; Kim, G.; Kim, J.; Qian, D.; Lee, S. Multi-Robot Path Planning Method Using Reinforcement Learning. *Appl. Sci.* **2019**, *9*, 3057. [CrossRef]
34. Zeng, J.; Ju, R.; Qin, L.; Hu, Y.; Yin, Q.; Hu, C. Navigation in Unknown Dynamic Environments Based on Deep Reinforcement Learning. *Sensors* **2019**, *19*, 3837. [CrossRef]
35. Pérez-Gil, Ó.; Barea, R.; López-Guillén, E.; Bergasa, L.M.; Gómez-Huélamo, C.; Gutiérrez, R.; Díaz-Díaz, A. Deep reinforcement learning based control for Autonomous Vehicles in CARLA. *Multimed. Tools Appl.* **2022**, *81*, 3553–3576. [CrossRef]
36. Ngo, B.-V.; Nguyen, T.-H.; Tran, D.-K.; Vo, D.-D. Control of a Smart Electric Wheelchair Based on EEG Signal and Graphical User Interface for Disabled People. In Proceedings of the 2021 International Conference on System Science and Engineering (ICSSE), Ho Chi Minh City, Vietnam, 26–28 August 2021; pp. 257–262.
37. Ngo, B.V.; Nguyen, T.H.; Nguyen, T.N. EEG Signal-Based Eye Blink Classifier Using Convolutional Neural Network for BCI Systems. In Proceedings of the 15th International Conference on Advanced Computing and Applications (ACOMP), Ho Chi Minh City, Vietnam, 24–26 November 2021; pp. 176–180.

38. Ngo, B.-V.; Nguyen, T.-H.; Ngo, V.-T.; Tran, D.-K.; Nguyen, T.-D. Wheelchair Navigation System using EEG Signal and 2D Map for Disabled and Elderly People. In Proceedings of the 5th International Conference on Green Technology and Sustainable Development (GTSD), Ho Chi Minh City, Vietnam, 27–28 November 2020; pp. 219–223.
39. Ngo, B.-V.; Nguyen, T.-H. Dense Feature-based Landmark Identification for Mobile Platform Localization. *Int. J. Comput. Sci. Netw. Secur.* **2018**, *18*, 186–200.
40. Nguyen, T.N.; Nguyen, T.H. Landmark-Based Robot Localization Using a Stereo Camera System. *Am. J. Signal Process.* **2015**, *5*, 40–50.
41. Wen, S.; Chen, X.; Ma, C.; Lam, H.K.; Hua, S. The Q-learning obstacle avoidance algorithm based on EKF-SLAM for NAO autonomous walking under unknown environments. *Robot. Auton. Syst.* **2015**, *72*, 29–36. [CrossRef]
42. Ryu, H.-Y.; Kwon, J.-S.; Lim, J.-H.; Kim, A.-H.; Baek, S.-J.; Kim, J.-W. Development of an Autonomous Driving Smart Wheelchair for the Physically Weak. *Appl. Sci.* **2022**, *12*, 377. [CrossRef]

Article

MuseStudio: Brain Activity Data Management Library for Low-Cost EEG Devices

Miguel Ángel Sánchez-Cifo *, Francisco Montero and María Teresa López

Laboratory of User Interaction & Software Engineering (LoUISE), Department of Computing Systems, University of Castilla-La Mancha, 02071 Albacete, Spain; francisco.msimarro@uclm.es (F.M.); maria.lbonal@uclm.es (M.T.L.)
* Correspondence: MiguelAngel.Sanchez@uclm.es

Citation: Sánchez-Cifo, M.A.; Montero, F.; López, M.T. MuseStudio: Brain Activity Data Management Library for Low-Cost EEG Devices. *Appl. Sci.* **2021**, *11*, 7644. https://doi.org/10.3390/app11167644

Academic Editors: Alexander E. Hramov, Hidenao Fukuyama and Jing Jin

Received: 26 May 2021
Accepted: 16 August 2021
Published: 20 August 2021

Publisher's Note: MDPI stays neutral with regard to jurisdictional claims in published maps and institutional affiliations.

Copyright: © 2021 by the authors. Licensee MDPI, Basel, Switzerland. This article is an open access article distributed under the terms and conditions of the Creative Commons Attribution (CC BY) license (https://creativecommons.org/licenses/by/4.0/).

Abstract: Collecting data allows researchers to store and analyze important information about activities, events, and situations. Gathering this information can also help us make decisions, control processes, and analyze what happens and when it happens. In fact, a scientific investigation is the way scientists use the scientific method to collect the data and evidence that they plan to analyze. Neuroscience and other related activities are set to collect their own big datasets, but to exploit their full potential, we need ways to standardize, integrate, and synthesize diverse types of data. Although the use of low-cost ElectroEncephaloGraphy (EEG) devices has increased, such as those whose price is below 300 USD, their role in neuroscience research activities has not been well supported; there are weaknesses in collecting the data and information. The primary objective of this paper was to describe a tool for data management and visualization, called MuseStudio, for low-cost devices; specifically, our tool is related to the Muse brain-sensing headband, a personal meditation assistant with additional possibilities. MuseStudio was developed in Python following the best practices in data analysis and is fully compatible with the Brain Imaging Data Structure (BIDS), which specifies how brain data must be managed. Our open-source tool can import and export data from Muse devices and allows viewing real-time brain data, and the BIDS exporting capabilities can be successfully validated following the available guidelines. Moreover, these and other functional and nonfunctional features were validated by involving five experts as validators through the DESMET method, and a latency analysis was also performed and discussed. The results of these validation activities were successful at collecting and managing electroencephalogram data.

Keywords: brain data; low-cost devices; EEG; BIDS; neuroscience; library

1. Introduction

Data are crucial elements of all systems that surround us today. Data collection is the process of gathering and measuring information on variables of interest, in an established systematic fashion that enables one to answer stated research questions, test hypotheses, and evaluate outcomes. The data collection component of research is common to all fields of study including physical and social sciences, humanities, business, etc. While methods vary by discipline, the emphasis on ensuring accurate and honest collection remains the same.

Indeed, storing valuable data is beneficial as this enables comparisons between different situations of the same subject, the same situation between different subjects, and a combination of both. As a result, proper treatment provides evidence in the scope of several environments, such as: patient monitoring with automatic health checks; sleep tracking with state detection; student performance analysis and prediction; obtaining a birds-eye view of how people travel, given the difficulties imposed by COVID-19; many other possibilities ruled by the quality of the data acquired.

In particular, the demand for ElectroEncephaloGraphy (EEG) and the devices that allow gathering brain activity has been increasing in the last few years. That interest is expected to keep growing in the future [1]. Medicine, marketing, interaction, and signal processing are some disciplines that require these kinds of products, especially those that feature dry sensors, knowing that some of them are relatively inexpensive.

Regardless of the field of study or preference for defining data (quantitative, qualitative), accurate data collection is essential to maintain the integrity of research. Both the selection of appropriate data collection instruments (existing, modified, or newly developed) and clearly delineated instructions for their correct use reduce the likelihood of errors occurring.

One of the main contributions is the compatibility with the Brain Imaging Data Structure (BIDS) [2] standard, which facilitates research activities related to the use of EEG devices. This standard allows researchers to organize and share the data associated with studies carried out in their laboratories. However, some EEG devices available on the market are not compatible with the BIDS. This issue makes managing recordings, sessions, and users a very difficult and inconvenient task. The majority of low-cost EEG devices have this limitation, and even though they are compatible with proprietary software for brain activity, the features included are limited and not very flexible [3].

In this context, we used a low-cost EEG device, known as Interaxon Muse 2 [4] (Muse and Muse S devices are also compatible). The manufacturer offered an SDK with computer support in the past (which was never compatible with Muse 2 and Muse S). However, it was deprecated, and currently, there is no viable alternative to use the devices in a research or professional environment. This only enables their connection to the original smartphone app, which is limited to guided meditation, and not intended for experiments.

To overcome the imposed limitations, we developed a Python library, called MuseStudio [5], that allows managing brain activity data from users with several sessions, including other helpful characteristics. The main research question that guided the development of this paper is the following: What (internal and external) features should a low-cost EEG library have to manage users' information while performing different activities? Among the solutions that MuseStudio provides, importing and exporting data stand as key differentiators using Muse. To ensure compatibility with current and future research, we focused on compliance with the best practices in data analysis and sharing [2]. Additionally, the recommendations from the OHBM COBIDAS MEEG committee [6] entirely apply to the introduced library in this paper.

There are multiple scenarios in which MuseStudio is helpful: sharing brain activity data recordings with colleagues thanks to the BIDS standard support; bulk importing other recordings in BIDS format, including raw recordings; converting to MNE format for further noise reduction, signal transformation, and analysis; viewing the experiments taking place in real time with several devices connected at the same time. For instance, a experiment can be performed with multiple Muse 2 devices, connected to a single computer running MuseStudio. Once the recording is finished, it can be converted for feature extraction and exported to share it with peers or attached to a research article for its publication, as it can be imported by anyone interested. Moreover, there is a big community around Muse devices due to its convenience and precision.

The article provides the related work, first. Then, the set of features included in the software with their specific purpose is presented. Afterwards, different examples of use are shown, outlining the results and the aspects of the visualization screen. Lastly, some insights about the necessity of this proposal are given, together with the discussion and the conclusion sections.

2. Background

An electroencephalogram is a data-intensive test that allows detecting abnormalities in brain waves, or the electrical activity of the brain [7]. During the procedure, electrodes consisting of small metal discs with thin wires are pasted onto the scalp. This technology has a wide variety of uses, especially in the emotion recognition domain, as the results of these articles showed [8,9].

Due to the previously annotated increasing demand of EEG devices, the number of available devices is on the rise, and they have many different characteristics [3,10,11]. Moreover, these devices are not only present in the research environment [12–17], but also in the entertainment one [18–20]. A recent article [21] analyzed the number of electrodes included in devices depending on their design. The authors concluded that the availability of more or less electrodes depends on the final application in which the device will be used. However, in these scenarios, the number of sensors is not the only key factor: data collection and software for supporting them are other relevant factors for success.

In our case, we focused on low-cost EEG devices [10,22,23]. The price requirement results in the number of electrodes being reduced, and then, the device has less capabilities depending on the field in which it is deployed [21]. Considering that there are still plenty of applications that can be explored and relate to meditation, relaxation, concentration, stress, and anxiety, many therapeutic and entertainment activities can be approached. In this research, Muse 2 was chosen among other viable alternatives. It features a sampling rate of 256 Hz for EEG concurrent signals, four dry capturing electrodes, plus frontal reference channels, an accelerometer, a gyroscope, a PhotoPlethysmoGraphy (PPG) sensor, a built-in battery, and Bluetooth. Following the 10-20 standard system, the device locates its sensors at AF7, AF8, TP9, and TP10.

Muse has been validated as a device for conducting Event-Related Potential (ERP) research [24]. This device has been compared with other wearable sensors resulting in high performance in the fields of ease of integration and applied usability [25]. In addition, many other studies have used Muse for several purposes, including brain wave activity detection during training [26], enjoyment evaluation [27], accelerometer measurement of head movement during surgery [28], and concentration and stress measurement during surgery [29].

In addition to the inherent hardware limitations of the devices, the software restrictions in terms of applications, software development kits, and application programming interfaces should be considered as well [3]. The great majority of software provided by manufacturers cannot manage activities, record sessions, and provide remote real-time visualization while participants are being evaluated. These issues are important limitations in supporting scientific activities. The community of users and researchers of, for instance, Muse products cannot perform data management for several sessions and different users and, later, analyze these data. The traditional manner of making evaluations with Muse is shown in Figure 1. To overcome some of these issues, MuseStudio allows storing data in a structured manner and sharing them.

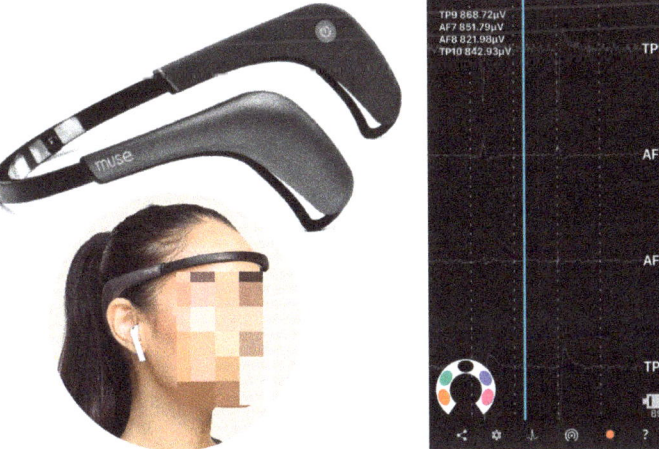

Figure 1. Traditional usage of Muse in experiments.

Specifically, Muse does not include brain data management software, nor real-time visualization, nor recording, so one cannot make use of its potential features. Our solution provides a Python library that allows working with those functionalities, even with several devices and places at the same time.

2.1. Connection with Muse Devices

In order to connect Muse with a computer using Bluetooth, there exist two applications that use Lab Streaming Layer (LSL) to transmit data. While MUSE-LSL [30] connects to one Muse, BlueMuse (https://github.com/kowalej/BlueMuse, accessed on 17 November 2020) can stream data from multiple Muse devices at the same time. However, note that the multidevice capabilities rely on the capacity of the receiving Bluetooth adapter. The data can be further recorded in files with LabRecorder (https://github.com/labstreaminglayer/App-LabRecorder, accessed on 30 November 2020), which can store data from several Muse devices in a single eXtensible Data Format (XDF) file.

As an alternative, there is a hardware-based framework [31] that measures EEG data obtained from 10 or more people using the Muse headband and allows acquiring EEG data at up to a 1 kHz frequency from up to 20 people simultaneously. However, in this hardware proposal, EEG data management cannot be provided, and it is only a graphical visualization tool.

The developed library requires some specific Python packages to work with the data, which are outlined in the repository. Additionally, it is compatible with other software applications that extend its functionality. The library has two main starting points: recordings already stored and live visualization of EEG data. The former requires files in XDF (https://github.com/sccn/xdf, accessed on 5 November 2020), which is a container specifically designed to include multichannel time series data with associated meta information. It can handle multiple types of data, including EEG. The latter adds compatibility with LSL (https://github.com/sccn/labstreaminglayer, accessed on 5 November 2020), which allows sending and receiving data in research experiments through the network. In addition, it features time synchronization and real-time data access in a structured manner. LSL can send several channels at the same time through the same stream, which ensures synchronization even at the channel level. As described previously, Muse has different kinds of data, including EEG, PPG, accelerometer, and gyroscope data. Those sensors do not function at the same sampling frequency, so they must be separated into different streams because of this incompatibility. The sampling frequencies are: 256 Hz for EEG, 64 Hz for PPG, 50 Hz for the accelerometer, and 50 Hz for the gyroscope. For this reason, the channels of the same type are sent in the form of a container with the captured data for a particular sample, but different types are sent over distinct containers. Four streams or containers are expected for a standard experiment with three channels in each of them, except for EEG, which contains four due to the four channels available. In general, equally sampled data are always sent in the same package.

2.2. Raw Data Import

MuseStudio facilitates the data import process from raw XDF files. Those can contain EEG, PPG, accelerometer, and gyroscope recordings from multiple Muses at the same time. That adds processing complexity because the captured data are not properly organized at recording time. The separation of recordings into different containers causes the reception of disordered data at the stream-type level. For instance, EEG and PPG may not be received in such an order, but it is ensured that channels inside those streams are correctly ordered and ready to use afterwards. For this reason, the library seeks the metadata of every channel to rearrange them into different sets of recordings, which can be further used accordingly. Figure 2 shows an example of a file containing the recording of two Muses (Step 1), which were used in a experiment simultaneously. The library then converts the file into four independent lists with the same length as the number of devices used (Step 2). Those lists already contain the information of the device used in each recording.

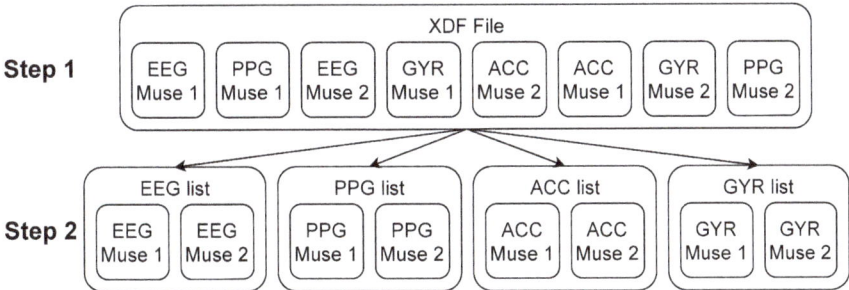

Figure 2. Example of XDF file conversion into separate lists.

Additionally, there is compatibility for importing all the XDF files located inside a particular folder. In such a case, the output remains the same, being four lists with a length that is equivalent to the sum of all the recordings inside all the files. Moreover, the library provides flexibility to researchers using Muse and Python because the data import approach only returns the lists without any other manipulation, so they can start working with native data.

2.3. Convert Data

Apart from being able to work with lists, there are two packages that are relevant exponents in their respective fields:

- MNE-Python [32] (or simply MNE) is an open-source package that allows the preprocessing, visualization, and analysis of human neurophysiological data;
- Pandas [33] provides high-level real-world data analysis and is becoming the most powerful and flexible open-source manipulation tool.

For researchers, being able to work with those packages is critical. This is especially relevant with MNE, because it is the most viable alternative when operating with EEG data. However, the package does not provide any kind of support for Muse, nor its native file formats. As a result, we provided a native implementation in MuseStudio that brings full interoperability for both packages.

Converting data into MNE format requires some considerations. In general, the conversion includes information about the sensor coordinates, the physiological coordinates of the study participants, the powerline frequency (which depends on the region, 50 Hz or 60 Hz), the data in volts, the channels' names, the associated annotations, and the type of data. In this case, only EEG data were considered because MNE does not work with PPG, accelerometer, and gyroscope data. The result of the transformation is an array of RawArray objects with the same order as the list obtained in the previous stage. Those objects can be iterated to perform the analysis in any research study.

The outcome of the conversion to Pandas is a list with several data frames that can be used for analysis using data science techniques. A single data frame has the following columns: timestamp, AF7, AF8, TP9, TP10, X_acc, Y_acc, Z_acc, X_gyr, Y_gyr, Z_gyr, 1_ppg, 2_ppg, 3_ppg. These correspond to all the streams provided by Muse. The differences between the sampling frequencies of the streams result in blank fields in rows.

2.4. BIDS Format Import and Export

MuseStudio, in order to support the data management of brain activity with Muse products, must consider data structural mechanisms. These mechanisms are inspired by the standard Brain Imaging Data Structure (BIDS) [2]. The addition of the BIDS support allows sharing Muse recordings with other researchers, even if they do not have the set of tools required to work with the device. The library, with the support of MNE-BIDS [34], manages all the necessary information to save the configuration data that the BIDS requires according to its specification. With that aim, two structures were defined, setup and participants, which are available in Appendix A. These are lists of dictionaries, so that every recording has a corresponding dictionary with its particular details.

The setup structure includes this information: the name of the subject; the session and run numbers; the acquisition parameters; the task performed; the processing label; the name of the recording; the coordinate space; the split of the continuous recording; the file name suffix and extension; the root path of the files. In the following examples, all the fields were simplified to None, but they should be modified accordingly.

The BIDS also requires a participants' file with the details of every member of the research study. The structure presented contains the information about the name of the subject (which must coincide with those in the setup), the age, the sex, the dominant hand, and root directory path. This file is an explicit recommendation of the BIDS specification [35], which suggests its addition in the root path of the main recordings directory.

The purpose of this configuration is to simplify how recordings are exported and imported. In addition, it allows knowing the characteristics of the experiments and the participants rapidly. When sharing one or more recordings, a researcher would only share the BIDS-formatted directory and the two updated lists described above. Using the designed method, no ambiguity is possible. Lastly, the creation of two structures, which are related thanks to the "subject" field (that is unique), avoids the repetition of information. A participant can have several recordings, but it is still the same participant. Therefore, his/her details must be added to the participants' structure only once, while the setup structure can hold several recordings.

2.5. Real-Time Remote View

Performing neural experiments usually requires real-time visualization of the brain signals captured. With time-based graphs, it is possible to detect how good the data received from the electrodes are, due to the fact that they may not have full contact with the skin and produce extra noise. Muse, with four electrodes, is especially vulnerable to this issue because one bad sensor can invalidate a full recording. The current available solution [30] only shows real-time visualization for one device at a time. Moreover, it only works on the same computer to which Muse is connected. This problem narrows down the flexibility when researchers want to perform experiments with multiple devices at once. The MuseStudio library provides access to real-time graphs no matter the number of devices attached. Additionally, it shows when the contact of the sensors with the skin is good for each of them independently.

Globalization has broken many barriers, and healthcare is one of them. Telemedicine [36] is increasingly being adopted for receiving medical treatment at a distance. In fact, patients who receive palliative care by telemedicine are very satisfied with the results. For this reason, we want everyone to be able to access neuroevaluations anywhere in the world without need to travel long distances to reach experts.

Instead of creating a local instance of a program, we created a web server with an IP address and a port that users can access through a web browser. This allows many users to be connected to the same endpoint, even if they are located outside the local area network. However, as a prerequisite, the server port must be connected to the Internet for external access. The implementation can be used straight away without authentication, and it is modular, so it can be integrated with other Python environments without adaptation, such as a website with a log-in required. The web browser must have JavaScript enabled to show the graphs. Finally, the complete set of options added is: sensor selection, update interval (from 200 ms to 5 s), play/pause, zoom in/out, and expand graphs.

3. Method

This section identifies and describes the internal characteristics of the MuseStudio library [5] available at https://github.com/miguelascifo/MuseStudio, accessed on 26 February 2021, which can be installed through the Python pip package manager (https://pypi.org/project/musestudio/, accessed on 3 March 2021) as well. The main internal requirements derived from the functionality of MuseStudio are presented in this section, including raw data importation and real-time data visualization. Raw data importation involves two activities: data conversion and data organization with the BIDS. All these activities are described below.

3.1. Data Conversion

MuseStudio can import from XDF files to work with Python arrays. There are two methods created for importing recordings, and both return the same data. The method *read_raw_xdf* handles one file, and *read_raw_xdf_dir* handles a directory with several XDF files. The following code shows an example of the latter:

```
stream_eeg, stream_acc, stream_ppg, stream_gyr, filenames = read_raw_xdf_dir("/path/to/directory")
```

where *stream_eeg*, *stream_acc*, *stream_ppg*, *stream_gyr*, and *filenames* are lists containing the data for EEG, the accelerometer, PPG, the gyroscope, and the file names of all recordings, respectively.

Once the data are imported using the methods exposed by the library, they can be converted into the MNE RawArray and Pandas data frame. One key difference between them is that MNE provides a powerful set of tools for EEG streams, but does not support the rest. For that reason, all data can be manipulated through data frames. Again, one method is necessary for the conversion:

```
raw = to_mne_eeg(eegstream = stream_eeg, line_freq = 50, filenames = filenames, nasion = [0,0,0],
   lpa = [0,0,0], rpa = [0,0,0])
```

where *eegstream* is the list of EEG data previously imported, *line_freq* the powerline frequency of the region (50 for Europe), and *filenames* the list of file names imported. The three following lists correspond to the nasion fiducial point (nasion), the left periauricular fiducial point (lpa), and the right periauricular fiducial point (rpa). Those indicate a precise reference for the EEG sensors' position on the head [37].

The conversion to the Pandas data frame gives the flexibility to import only the streams in which the researcher is interested. This example of usage includes all the streams at once:

```
df = to_df(mne_eeg = raw, eegstream = stream_eeg, accstream = stream_acc, ppgstream = stream_ppg,
   gyrstream = stream_gyr)
```

the parameters being those variables that were already described. The resulting data frame contains blank spaces (Pandas NotaNumber data types) between rows in the last three columns. That happens because the sampling rates are different, as explained previously.

3.2. Working with the BIDS Specification

The BIDS specification establishes the directory structure to standardize how researchers store and share EEG recordings. The huge advantages make using it useful for working in a collaborative environment. To simplify the process, we created the *setup* structure. In order to export recordings, the BIDS file name paths are necessary, which is the information of the type BIDSPath object. Afterwards, the paths of the recordings included in *setup* are returned. Then, the first step is to execute the appropriate method provided for such a task:

```
bids_paths = create_bids_path(setup = setup)
```

After that, there is another method that uses those paths together with other parameters to finally export the recordings in the BIDS format. This example uses the *participants* structure to specify the characteristics of the subjects and the list of BIDSPath objects:

```
export_bids(raweeg = raw, bids_paths = bids, participants = participants, overwrite = False,
   verbose = False)
```

With that process, the recordings are exported to the directory indicated in the "root" field inside *setup*. In contrast, importing from the BIDS requires executing one method with a single parameter, which is *setup*. It returns the list of RawArray objects and the list of BIDSPath objects for the recordings in *setup*:

```
raw, bids_paths = import_bids(setup = setup)
```

With the solution proposed, anyone can import directly into MNE to start working with Muse as if it were any other more advanced device, provided that other researchers have exported the recordings previously.

3.3. Signal Visualization

Performing experiments with EEG can be very complex due to the difficulty of creating high-quality recordings. One of the main issues, apart from the design of the experiment itself, is measuring how well the data were captured. Electrodes inside devices are very sensitive to electromagnetic noise, so ensuring good skin contact is critical to avoid inconsistent results across recordings. For this reason, the library includes the necessary features to enable researchers to watch the signals of several devices in real time.

Sometimes, experiments are not run by medical experts, which is especially the case for low-cost devices. Therefore, we ensured that anyone can have access to the data while participants are being evaluated. The web server is started from the machine to which the devices are connected and returns to the console the internal IP address, together with the associated port. The process was simplified as two methods:

```
start_streaming(search_streams(), debug=True)
```

There are two different ways of visualizing the signals, compressed and expanded. In Figure 3, the overall compressed view of the website is shown with two devices at the same time. The latter is shown in Figure 4. Additionally, there are controls for the update interval (from 200 ms to 5 s), the channels to watch, the zoom level, and the possibility to play and pause the live view. For anyone without deep knowledge about performing experiments, we included a marker to know if a particular electrode had good skin contact. This helps to keep noise sufficiently low to retain a high probability of success. Lastly, the library automatically detects how many Muses are connected to the computer and adapts the interface to show those.

In Figure 5, we show our solution for experiments with multiple devices connected at the same time, while watching the streams. Additionally, the data file structure exported using the BIDS format is presented.

Figure 3. Overview of the interface.

Figure 4. Expanded view of the signals.

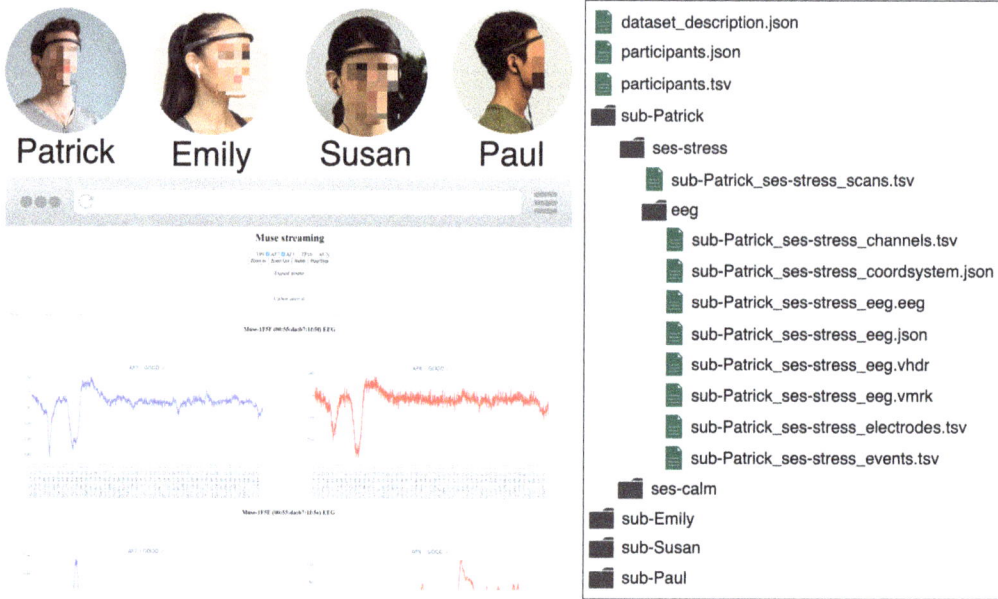

Figure 5. Solution designed for experiments.

4. Validation

The main validation activities of MuseStudio are described in this section. The relevant external features of MuseStudio were evaluated by using qualitative and quantitative methods. For the evaluation of the functional and nonfunctional features of a library for supporting brain data management, such as MuseStudio, a well-known evaluation method from the software engineering field was use. Performance and latency are other important elements when brain data are collected and visualized.

First, over the years, many software engineering methodology evaluation frameworks have been published. DESMET [38] is a methodology for evaluating software engineering methods/tools by Barbara Kitchenham. DESMET can be used to compare a generic method or a method that is a specific approach within a generic method or tool. This methodology has been used in other articles for evaluation purposes [39]. According to DESMET, there are two types of evaluations:

1. The evaluation of the measureable effects of using a method or tool;
2. The evaluation of the appropriateness of the method or tool, i.e., how usable or useful the method is.

DESMET refers to the measureable effects of using a method as quantitative or objective, while method appropriateness is referred to as qualitative, feature analysis, or subjective. Method appropriateness is accessed usually in terms of features provided by the method/tool or the training requirements. Another important consideration is how to organize the evaluation process. According to DESMET, for a qualitative evaluation, it can be organized as a survey, a case study, or a formal experiment. In qualitative screening, it can be organized as a feature screening mode, a survey, a formal experiment, or a case study. DESMET qualitative feature screening mode can be performed by a single person for a number of methods where the evaluator not only determines the features to be accessed and their rating scale, but also performs the assessment. In qualitative screening mode, the evaluations are usually based on the literature describing the software method, rather than actually using the method.

Secondly, a latency study in MuseStudio was also conducted, and the data gathered are shown.

4.1. Analyzing the Main Features of MuseStudio

Using as the input a demonstration of MuseStudio, the evaluation of this library was carried out using DESMET [38]. This is a set of techniques applicable to evaluating both software engineering methods and tools. We used the method based on a qualitative case study, which describes a feature-based evaluation. Following the guidelines specified for this technique, an initial list of features that a library or tool for EEG data management should provide was defined (see Table 1). These features were established by two experts (full professors) in cognitive neuropsychology from the University of Castilla-La Mancha (UCLM). As can be observed, some of the features are directly related to the availability of the BIDS.

DESMET was deployed by involving five experts. First, two experts were asked about the main requirements a library for low-cost EEG devices should provide. Second, another three experts were involved to validate MuseStudio by considering the previously proposed requirements. All the experts were professionals with knowledge and skills related to EEG devices, neuroscience, and psychology.

Once Table 1 has been filled in by the experts, DESMET determines the importance degree that should be assigned to each identified feature. Specifically, the importance degrees are Mandatory (M), Highly Desirable (HD), Desirable (D), and Nice to have (N). This importance was also established by the consulted experts.

By using these importance degrees, Table 2 was filled in. As can be noticed, the most important functional and nonfunctional requirements to be supported are signal visualization, import and export data management, and scalability.

Table 1. List of features for MuseStudio's evaluation.

Feature	Description
Signal visualization	The tool should be able to provide graphic visualization of the associated signals to each sensor of the headband and differentiate among them.
Session management (import)	The tool has to store or import the associated data of each session and use and differentiate among them.
Session management (export)	The tool has to allow sharing stored data, that is the tool should be able to export the stored data of each session and user.
User control	The tool has to provide user control during a session. For instance, the graphical visualization of EEG signals should be stopped and restarted.
Scenario identification	The tool should be able to identify rare scenarios, for instance a poorly worn headband.
Easy of data reviewing	The stored data of each session should be easy to review and manage.
Consistency	The stored data of each session and user should be jointly managed.
Real time (same time)	The tool must allow following a session in real time, including minimal latency to improve the performance.
At a distance (different place)	The tool must allow following a session at a distance, so that the user/headband can be in different places and the data visualization can be performed in different places.
Scalability	The tool must allow using several headbands simultaneously with different users.
Guided user interface	The tool must provide a user-friendly interface to operate easily with its features.

Table 2. Relevance of features (Mandatory (M), Highly Desirable (HD), Desirable (D), and Nice to have (N)).

Feature	Importance
Signal visualization	M
Session management (import)	M
Session management (export)	M
User control	HD
Scenario identification	HD
Easy for data reviewing	D
Consistency	D
Real time (same time)	HD
At a distance (different place)	HD
Scalability	M
Guided user interface	HD

Afterwards, according to DESMET, a scale to evaluate each of the described features should be provided. The scale proposed by DESMET (see Table 3) was applied to evaluate each feature according to the following factors: Conformance Acceptability Threshold (CAT) and Conformance score obtained (CSO) for MuseStudio. In particular, three experts (associate professors) from the University of Castilla-La Mancha with experience in the fields emotion recognition, health psychology, and signal processing/computer science agreed about the values of CSO_i.

Table 3. Judgment scale to assess tool support for a feature.

Generic Scale Point	Definition of Scale Point	Scale Point Mapping
Makes things worse	Causes confusion. The way the feature is represented makes its modeling difficult and/or encourages its incorrect use.	−1
No support	Fails to recognize it. The approach is not able to model a certain feature.	0
Little support	The feature is supported indirectly, for example using another model/approach in a nonstandard combination.	1
Some support	The feature is explicitly in the feature list of the model. However, it does not cater to some aspects of the feature use.	2
Strong support	The feature is explicitly in the feature list of the model. All aspects of the feature are covered, but its use depends on the expertise of the user.	3
Very strong support	The feature is explicitly in the feature list of the model. All aspects of the feature are covered, and the approach provides a guide to assist the user.	4
Full support	The feature appears explicitly in the feature list of the model. All its aspects are covered, and the approach provides a methodology to assist the user.	5

Once each feature was evaluated, the difference between the CAT and CSO factors was computed as shown in the column Difference (Dif) in Table 4.

Therefore, in order to interpret the values shown in Table 4, the following equations should be considered:

$$Imp_i = \text{Level of relevance of each feature (i)}$$

$$CAT_i = \text{Level of support of each feature (i)}$$

$$CSO_i = \text{Quantitative evaluation of each feature (i) by specialists in several fields}$$

$$Dif_i = CSO_i - CAT_i \tag{1}$$

$$Score_i = Imp_i * Dif_i \tag{2}$$

$$Total = \sum_{i=1}^{features} Score_i \tag{3}$$

We should highlight that a variation of the DESMET method was created. The Importance (Imp) of each feature was weighed using a scale from 1 to 4 (Nice to have—1, Desirable—2, Highly Desirable—3, Mandatory—4). The importance was used to compute the final score of each feature or requirement by multiplying the importance by the difference. This computation is shown in the column Score (Sco) in Table 4. This score is useful for comparing different alternatives, but in our case, the score was only for the MuseStudio's valorization. Lastly, the final score of each technique (Total) was obtained by adding the scores of all the features.

The MuseStudio library achieved a positive total score (15 points). Moreover, it was especially evaluated positively for the "at a distance" feature, since MuseStudio provides full support for exporting the brain activity data. It was also highlighted that the MuseStudio tool has consistency and easily represents the requirements' importance, giving no support to determining which requirements are more important than the others. In any case, MuseStudio provides facilities for data gathering and collection in conformance with the BIDS proposal. Brain data from Muse devices are organized and structured with MuseStudio, and these data can be visualized, imported, exported, and analyzed.

Table 4. Results of MuseStudio's evaluation.

Feature	Imp	CAT	CSO	Dif	Score
Signal visualization	4	5	5	0	0
Session management (import)	4	5	5	0	0
Session management (export)	4	4	5	1	4
User control	3	3	4	1	3
Scenario identification	3	3	3	0	0
Easy for data reviewing	2	2	4	2	4
Consistency	2	2	4	2	4
Real time (same time)	3	3	4	1	3
At a distance (different place)	3	3	5	2	6
Scalability	4	4	4	0	0
Guided user interface	3	3	0	−3	−9
Total					**15**

In addition, as DESMET suggests, we performed a comparison of the percentage of each feature satisfied by MuseStudio. Figure 6 illustrates the results relative to the considered features. The outcomes of the validation are graphically shown in Figure 6. All previously established requirements were fully achieved. However, additional effort could be made on the user interface feature. At this moment, the information of the sessions and participants must be established directly by modifying this information in different files. Forms may be designed to ease these tasks.

Understanding the score requires knowing how DESMET works. First, the level of importance of a feature was determined by experts without trying the library (between −1 and 5). Thereafter, other experts determined how well implemented a particular feature was (between −1 and 5 again).

The current implementation of MuseStudio satisfies the requirements or features related to visualization, import data, scenario identification, and scalability. Other features of MuseStudio, such as data reviewing and data consistency, are more than satisfied, and the rest are also oversatisfied. At this time, the identified weakness of MuseStudio is that its users need to have certain knowledge about Python, because it does not have a guided user interface yet.

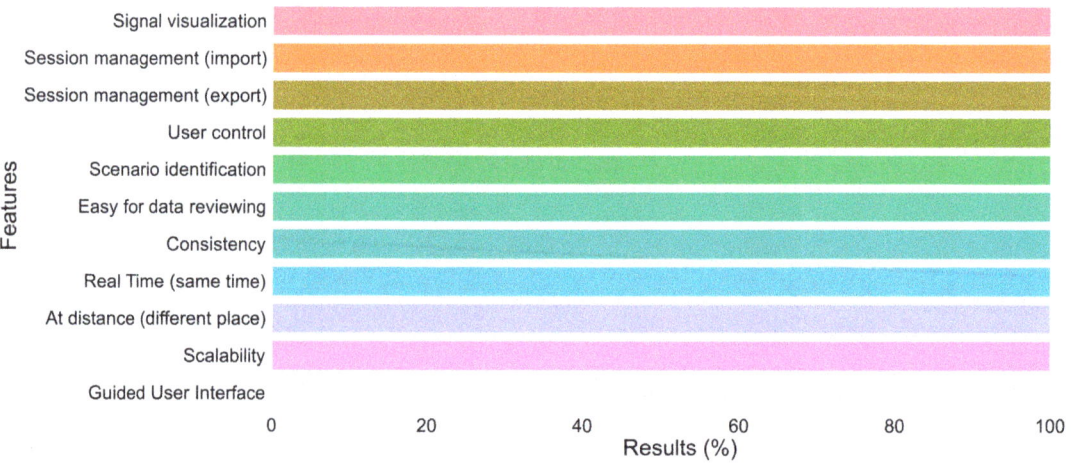

Figure 6. Results depending on each feature.

4.2. Latency Test

Some experiments with MuseStudio may require real-time data visualization, which is an included feature in the library. However, researchers may have special requirements in terms of the latency between the time an event occurs in the brain of a participant and the moment it is visible on screen. For this reason, we performed a latency test with all the different update intervals selectable. Those intervals were: 200 ms, 500 ms, 750 ms, 1 s, 1.5 s, 2 s, 3 s, and 5 s.

The design of the experiment measured the latency with real events, having a subject wearing Muse and a computer with the device connected. In particular, the device is able to capture eye blinks clearly, so this was the event that was going to be recorded repeatedly with the slow-motion camera of a Samsung Galaxy S20+ (Sony IMX555 main camera sensor) at a resolution of 1920 × 1080 and 240 frames per second. Then, the procedure consisted of a slow-motion camera pointing at the screen showing the real-time graphs and the subject performing the experiment, simultaneously. Afterwards, the subject was instructed to blink his/her eyes exactly when the graph updated. We are aware that there might be a slight variability regarding the time at which the subject blinks, so the experiment was repeated ten times with all the intervals, and then, we calculated the arithmetic mean between the values. Figure 7 shows a summary of the recording stage of the experiment. When that phase was finished, we loaded the video into an editor to count the frames between the blinks and the instant of those shown on screen. Once the frames were collected, we converted them into seconds knowing that 240 frames is equivalent to 1 s.

For the sake of reproducibility, Muse was connected to a computer with these specifications: Intel Core i7-9750H (base frequency 2.60 GHz and turbo frequency 4.50 GHz), 16 GB of RAM, and SSD (although no brain data were stored). The screen had an input lag of 5ms, which was discounted to each measurement. The connection with another computer to the server was not contemplated because that would add the latency of the network. Time synchronization was ensured by the LSL protocol [40], which achieves sub-millisecond accuracy on a local network without further action on practically all consumer PC hardware. The results are presented in Figure 8 through a bar plot that includes the variability of the measurements for each interval. It is observable that update intervals equal to or greater than one second showed the events with the correct timing and the expected latency. However, less than one-second values did not show a latency equivalent to the interval. This happened due to a combination of two different sources of delay: the time it takes for the device to send data and the time needed for the computer to attach the new values, create a visual representation, and update the interface. The difference in latency between those values was around 1ms, which did not correspond to the interval chosen. Nevertheless, we maintained those options because higher-performing CPUs are able to reduce the latency tested.

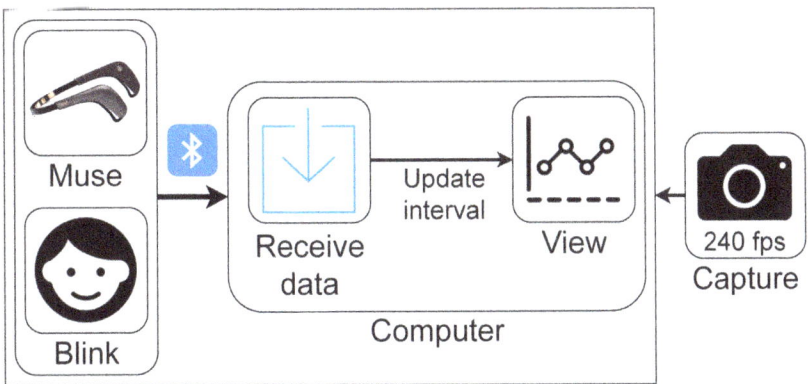

Figure 7. Design of the setup for the latency test.

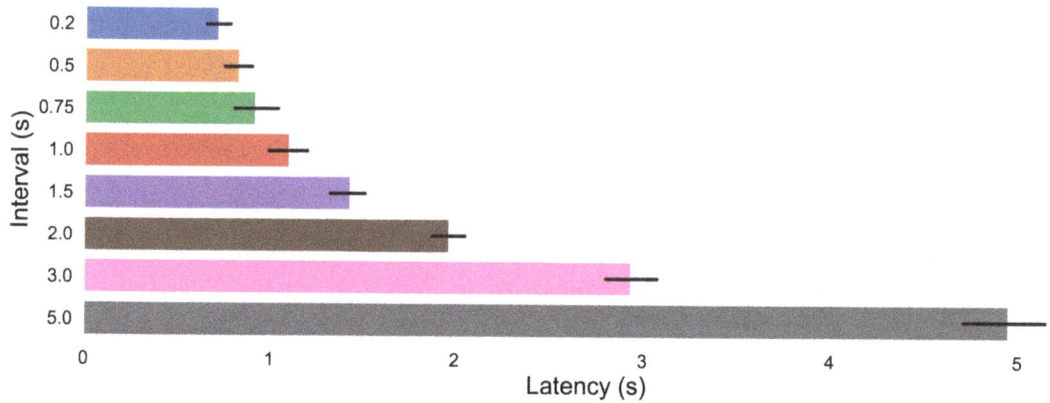

Figure 8. Results of the latency test.

5. Discussion

The objective of the study covered the creation of an open-source software product that allows working with brain activity data and facilitates the management of activities designed for performing experiments. In particular, Muse was chosen as the low-cost device to allow researchers to focus on their research.

The library MuseStudio provides a set of tools for management activities, including the import, conversion, export, and visualization of brain data. Thus, the solution adapts to real-time usage and recorded experiments. Moreover, those steps can be performed far from the place where the trial is being conducted, due to the tools provided.

The internal features of MuseStudio are the following: open-source cross-platform library; developed for Python 3 [41]; complies with the best practices in data analysis [2] and the recommendations from the OHBM COBIDAS MEEG committee [6]; allows visualizing real-time data from multiple devices concurrently without being in the same place; imports data from unlimited raw recordings and multiple devices in a structured manner; exports using the standard for EEG data; converts to MNE- and Pandas-compatible data formats. These internal features drove the MuseStudio development activities. Moreover, other external features were identified by two external experts in neuroscience.

Making the library open-source allows its usage and modification without worries, so other researchers and people interested in this field can use low-cost and minimally invasive devices in their experiments. In addition, the community can help by introducing new features and adapt the library to their particular necessities. It has been developed for all three major operating systems (Windows, Linux, and macOS) to ensure compatibility. As a prerequisite to use the library, having prior knowledge of Python is required. Python has converted into the preferred programming language for data science [41].

MuseStudio complies and follows the recommendations provided by the BIDS standard for neuroscience [42] to manage data recordings adequately. Therefore, it can import and export the data associated with multiple subjects and sessions using multiple devices. These data are not limited to the tasks that Muse natively supports, such as meditation. Instead, it supports any other validated activity. Following the BIDS [43] standard allows sharing data between partners and replicating experiments easily through the import and export functionalities.

The external features of MuseStudio were validated by three external experts in neuroscience. They validated the presence of these features and their relevance. All these features, initially established by using the DESMET method, were identified and properly evaluated in the current version of MuseStudio.

In MuseStudio, there are no limitations softwarewise, except for the lack of a guided user interface. This software shortcoming was previously identified and discussed. It can be overcome by designing and integrating user interface forms to provide session and participant descriptors and identifiers. Hardwarewise, the number of Muse devices simultaneously connected to a single computer is limited by the bandwidth and throughput of the Bluetooth module, which is different across machines. The library supports pausing the visualization at some point to explore a certain moment in time, and if the connection is lost, it automatically continues after reconnection. Additionally, the latency was tested with real-world usage in a controlled environment to maximize the delay between an event and its visualization on screen. The results showed exact timing from a 1 s update interval and times that varied depending on the interval if it was lower than 1 s. Nonetheless, those can be further reduced using a computer with better specifications.

In summary, MuseStudio shows that low-cost devices related to neuroscience, such as Muse, can have a complete set of tools to manage brain data. It offers features that increase flexibility, reliability, and the ease of data management.

6. Conclusions and Further Work

An electroencephalogram is an electrophysiological monitoring method that records the electrical activity of the brain. It is a noninvasive technique through electrodes placed on the scalp, and therefore, it is suitable for use in a wide variety of situations, not just the laboratory ones. Moreover, this method is data intensive, and in order to successfully manage these data, effective data visualization and collection are important. Software applications are needed for brain data management.

The article had special interest in affordable and low-cost EEG devices. A particular one is Muse from Interaxon, which although limited by the number of electrodes, is widely used for meditation and relaxation activities [14,15,17], being useful in the contexts of stress and anxiety. In this paper, we wanted to identify internal and external features for EEG data management and low-cost EEG devices; this collection of features should be the answer to our research question. These requirements were proposed and identified in the Method and Validation sections of this paper. In the internal dimension, several requirements were proposed, data import and conversion, the BIDS management of data, and real-time data visualization, and all these features were considered in the MuseStudio implementation. Later, using DESMET, external requirements were proposed and used in a validation activity. These external features were related to session data management (data importation and exportation), data visualization (signal visualization, consistency, scenario identification, easy for data reviewing), and ease of operation (scalability, same time, and different place).

Nevertheless, the software associated (manufacturer developed) with this device has many limitations, due to the lack of support for data collection and management. In this article, we overcame this deficiency with the creation of a library to manage brain activity data using Muse (different versions of Muse, Muse 2 and Muse S). MuseStudio provides a set of tools that facilitate storing, importing, exporting, visualizing, and sharing data. This article described the main features and strengths of the library, as well as a validation of those features, including to what extent they were achieved. In terms of hardware limitations, they were set by the particular low-cost device, Muse in this case. Depending on the specifications, some domains may be out of scope, not providing valuable insights.

Initially, several experts from the Psychology Department of the University of Castilla-La Mancha helped to determine which were the functional and nonfunctional features that a library related to brain data should include. Thanks to this collaboration, a set of features was identified by these experts to determine what tasks a software brain data management software tool should be able to perform. These features were used in order to validate MuseStudio by other experts, but additionally, these features can be used to compare MuseStudio with other alternatives in the future. In our functional and nonfunctional validation, other experts identified the presence or absence of those features using surveys, heuristic evaluation techniques, and analyzing MuseStudio in particular.

The library implemented is already a relevant contribution because it covers the initial necessities established. This library has been shared with the community through an open-source license [5]. Since its inception, MuseStudio has not been intended for the general public, but rather for researchers who are already familiar with the use and interpretation of brain signals. However, we can address other evaluations in the future as the library grows and improves. For instance, it could be useful as soon as a graphical interface is included, which is the main nonfunctional limitation. This feature would encourage the use of the library.

The library can be further improved by adding authentication and additional security capabilities. At this moment, for instance, the authentication of users and sessions must be performed by analysts, and these identification activities are not supported by the current version of MuseStudio. In this sense, users that need remote access should be able to establish secure connections between peers.

Author Contributions: Conceptualization, M.Á.S.-C. and F.M.; methodology, M.Á.S.-C., F.M. and M.T.L.; validation, M.Á.S.-C., F.M. and M.T.L.; investigation, M.Á.S.-C.; resources, M.T.L.; writing—original draft preparation, M.Á.S.-C.; writing—review and editing, F.M. and M.T.L.; funding acquisition, M.T.L. and F.M. All authors have read and agreed to the published version of the manuscript.

Funding: This work was supported by the European Regional Development Fund under the Grant Evaluando la eXperiencia de Usuario de personas mayores con técnicas de Neuroevaluación—NeUX (SBPLY/17/180501/000192).

Data Availability Statement: Not Applicable.

Acknowledgments: The authors acknowledge the participation of the staff from the Psychology Department and the Computer Science Department at the University of Castilla-La Mancha and the Spanish Ministerio de Economía y Competitividad for partially supporting this research through Grant 2gether (PID2019-108915RB-I00).

Conflicts of Interest: The authors declare no conflict of interest.

Abbreviations

The following abbreviations are used in this manuscript:

EEG   ElectroEncephaloGraphy
PPG   PhotoPletismoGraphy
BIDS Brain Imaging Data Structure
LSL   Lab Streaming Layer
XDF   eXtensible Data Format
ERP   Event-Related Potential

Appendix A. Structures Defined for BIDS

```
setup = [
{
"subject": None,
"session": None,
"task": None,
"acquisition": None,
"run": None,
"processing": None,
"recording": None,
"space": None,
"split": None,
"root": None,
"suffix": None,
"extension": None
}
```

```
                    ]

            participants = [
             {
              "subject": None,
              "age": None,
              "sex": None,
              "hand": None,
              "root": None
             }
            ]
```

References

1. TechNavio. *Global EEG Electrodes Market 2019–2023*; Technical Report; TechNavio: Elmhurst, IL, USA, 2019.
2. Pernet, C.; Garrido, M.; Gramfort, A.; Maurits, N.; Michel, C.; Pang, E.; Salmelin, R.; Schoffelen, J.M.; Valdes-Sosa, P.; Puce, A. Best Practices in Data Analysis and Sharing in Neuroimaging using MEEG. *Preprint* **2018**. [CrossRef]
3. Soufineyestani, M.; Dowling, D.; Khan, A. Electroencephalography (EEG) Technology Applications and Available Devices. *Appl. Sci.* **2020**, *10*, 7453. [CrossRef]
4. InteraXon. *Muse—Meditation Made Easy with the Muse Headband*; InteraXon: Toronto, ON, Canada, 2021.
5. Sanchez-Cifo, M.A.; Montero, F.; López, M.T. MuseStudio. *Zenodo* **2021**. [CrossRef]
6. Pernet, C.; Garrido, M.I.; Gramfort, A.; Maurits, N.; Michel, C.M.; Pang, E.; Salmelin, R.; Schoffelen, J.M.; Valdes-Sosa, P.A.; Puce, A. Issues and recommendations from the OHBM COBIDAS MEEG committee for reproducible EEG and MEG research. *Nat. Neurosci.* **2020**, *23*, 1473–1483. [CrossRef] [PubMed]
7. Johns Hopkins University. *Electroencephalogram (EEG)*; Johns Hopkins University: Baltimore, MD, USA, 2021.
8. Wang, L.; Liu, H.; Zhou, T.; Liang, W.; Shan, M. Multidimensional emotion recognition based on semantic analysis of biomedical eeg signal for knowledge discovery in psychological healthcare. *Appl. Sci.* **2021**, *11*, 1338. [CrossRef]
9. Al-Nafjan, A.; Hosny, M.; Al-Ohali, Y.; Al-Wabil, A. Review and classification of emotion recognition based on EEG brain-computer interface system research: A systematic review. *Appl. Sci.* **2017**, *7*, 1239. [CrossRef]
10. Peake, J.M.; Kerr, G.; Sullivan, J.P. A Critical Review of Consumer Wearables, Mobile Applications, and Equipment for Providing Biofeedback, Monitoring Stress, and Sleep in Physically Active Populations. *Front. Physiol.* **2018**, *9*, 743. [CrossRef]
11. LaRocco, J.; Le, M.D.; Paeng, D.G. A Systemic Review of Available Low-Cost EEG Headsets Used for Drowsiness Detection. *Front. Neuroinform.* **2020**, *14*, 42. [CrossRef]
12. Alsuradi, H.; Park, W.; Eid, M. EEG-Based Neurohaptics Research: A Literature Review. *IEEE Access* **2020**, *8*, 49313–49328. [CrossRef]
13. Lai, C.Q.; Ibrahim, H.; Abdullah, M.Z.; Abdullah, J.M.; Suandi, S.A.; Azman, A. Literature survey on applications of electroencephalography (EEG). In *AIP Conference Proceedings*; American Institute of Physics Inc.: Melville, NY, USA, 2018; Volume 2016. [CrossRef]
14. Millstine, D.M.; Bhagra, A.; Jenkins, S.M.; Croghan, I.T.; Stan, D.L.; Boughey, J.C.; Nguyen, M.D.T.; Pruthi, S. Use of a Wearable EEG Headband as a Meditation Device for Women With Newly Diagnosed Breast Cancer: A Randomized Controlled Trial. *Integr. Cancer Ther.* **2019**, *18*, 1534735419878770. [CrossRef] PMID:31566031.
15. O'Sullivan, M.; Temko, A.; Bocchino, A.; O'Mahony, C.; Boylan, G.; Popovici, E. Analysis of a Low-Cost EEG Monitoring System and Dry Electrodes toward Clinical Use in the Neonatal ICU. *Sensors* **2019**, *19*, 2637. [CrossRef]
16. Segawa, J.A. Hands-on Undergraduate Experiences Using Low-Cost Electroencephalography (EEG) Devices. *J. Undergrad. Neurosci. Educ. JUNE A Publ. FUN Fac. Undergrad. Neurosci.* **2019**, *17*, A119–A124.
17. Giorgi, A.; Ronca, V.; Vozzi, A.; Sciaraffa, N.; di Florio, A.; Tamborra, L.; Simonetti, I.; Aricò, P.; Di Flumeri, G.; Rossi, D.; et al. Wearable Technologies for Mental Workload, Stress, and Emotional State Assessment during Working-Like Tasks: A Comparison with Laboratory Technologies. *Sensors* **2021**, *21*, 2332. [CrossRef]
18. Vasiljevic, G.A.M.; de Miranda, L.C. Brain–Computer Interface Games Based on Consumer-Grade EEG Devices: A Systematic Literature Review. *Int. J. Hum.-Comput. Interact.* **2020**, *36*, 105–142. [CrossRef]
19. Svetlov, A.S.; Nelson, M.M.; Antonenko, P.D.; McNamara, J.P.; Bussing, R. Commercial mindfulness aid does not aid short-term stress reduction compared to unassisted relaxation. *Heliyon* **2019**, *5*, e01351. [CrossRef]
20. Stockman, C. Can a Technology Teach Meditation? Experiencing the EEG Headband InteraXon Muse as a Meditation Guide. *Int. J. Emerg. Technol. Learn.* **2020**, *15*, 83. [CrossRef]
21. Park, S.; Han, C.H.; Im, C.H. Design of Wearable EEG Devices Specialized for Passive Brain–Computer Interface Applications. *Sensors* **2020**, *20*, 4572. [CrossRef]

22. Rieiro, H.; Diaz-Piedra, C.; Morales, J.M.; Catena, A.; Romero, S.; Roca-Gonzalez, J.; Fuentes, L.J.; Di Stasi, L.L. Validation of Electroencephalographic Recordings Obtained with a Consumer-Grade, Single Dry Electrode, Low-Cost Device: A Comparative Study. *Sensors* **2019**, *19*, 2808. [CrossRef] [PubMed]
23. Peterson, V.; Galván, C.; Hernández, H.; Spies, R. A feasibility study of a complete low-cost consumer-grade brain-computer interface system. *Heliyon* **2020**, *6*, e03425. [CrossRef] [PubMed]
24. Krigolson, O.E.; Williams, C.C.; Norton, A.; Hassall, C.D.; Colino, F.L. Choosing MUSE: Validation of a Low-Cost, Portable EEG System for ERP Research. *Front. Neurosci.* **2017**, *11*, 109. [CrossRef] [PubMed]
25. Leape, C.; Fong, A.; Ratwani, R.M. Heuristic usability evaluation of wearable mental state monitoring sensors for healthcare environments. In *Proceedings of the Human Factors and Ergonomics Society*; Human Factors an Ergonomics Society Inc.: Washington, DC, USA, 2016; pp. 583–587. [CrossRef]
26. Chen, H.J.; Lin, C.J.; Lin, P.H.; Guo, Z.H. The effects of 3d and 2d imaging on brain wave activity in laparoscopic training. *Appl. Sci.* **2021**, *11*, 862. [CrossRef]
27. Abujelala, M.; Sharma, A.; Abellanoza, C.; Makedon, F. Brain-EE: Brain enjoyment evaluation using commercial EEG headband. In Proceedings of the 9th ACM International Conference on PErvasive Technologies Related to Assistive Environments (PETRA '16), Corfu Island, Greece, 29 June–1 July 2016; ACM: New York, NY, USA, 2016; Volume 2016. [CrossRef]
28. Viriyasiripong, S.; Lopez, A.; Mandava, S.H.; Lai, W.R.; Mitchell, G.C.; Boonjindasup, A.; Powers, M.K.; Silberstein, J.L.; Lee, B.R. Accelerometer measurement of head movement during laparoscopic surgery as a tool to evaluate skill development of surgeons. *J. Surg. Educ.* **2016**, *73*, 589–594. [CrossRef]
29. Maddox, M.M.; Lopez, A.; Mandava, S.H.; Boonjindasup, A.; Viriyasiripong, S.; Silberstein, J.L.; Lee, B.R. Electroencephalographic monitoring of brain wave activity during laparoscopic surgical simulation to measure surgeon concentration and stress: Can the student become the master? *J. Endourol.* **2015**, *29*, 1329–1333. [CrossRef]
30. Barachant, A.; Morrison, D.; Banville, H.; Kowaleski, J.; Shaked, U.; Chevallier, S.; Torre Tresols, J.J. muse-lsl. *Zenodo* **2019**. [CrossRef]
31. Lee, S.; Cho, H.; Kim, K.; Jun, S.C. Simultaneous EEG Acquisition System for Multiple Users: Development and Related Issues. *Sensors* **2019**, *19*, 4592. [CrossRef]
32. Gramfort, A.; Luessi, M.; Larson, E.; Engemann, D.A.; Strohmeier, D.; Brodbeck, C.; Goj, R.; Jas, M.; Brooks, T.; Parkkonen, L.; Hämäläinen, M. MEG and EEG data analysis with MNE-Python. *Front. Neurosci.* **2013**, *7*, 267. [CrossRef]
33. Pandas. The pandas development team. *Zenodo* **2020**. [CrossRef]
34. Appelhoff, S.; Sanderson, M.; Brooks, T.L.; van Vliet, M.; Quentin, R.; Holdgraf, C.; Chaumon, M.; Mikulan, E.; Tavabi, K.; Höchenberger, R.; et al. MNE-BIDS: Organizing electrophysiological data into the BIDS format and facilitating their analysis. *J. Open Source Softw.* **2019**, *4*. [CrossRef]
35. BIDS-Contributors. The Brain Imaging Data Structure (BIDS) Specification. *Zenodo* **2021**, 25–26. [CrossRef]
36. Calton, B.; Abedini, N.; Fratkin, M. Telemedicine in the Time of Coronavirus. *J. Pain Symptom Manag.* **2020**, *60*, e12–e14. [CrossRef]
37. Martínez, E.E.G.; González-Mitjans, A.; Bringas-Vega, M.L.; Valdés-Sosa, P.A. Automatic detection of fiducials landmarks toward development of an application for EEG electrodes location (digitization): Occipital structured sensor based-work. *arXiv* **2019**, arXiv:1912.07221.
38. Kitchenham, B.A. Evaluating Software Engineering Methods and Tool Part 1: The Evaluation Context and Evaluation Methods. *SIGSOFT Softw. Eng. Notes* **1996**, *21*, 11–14. [CrossRef]
39. F., I.D.; E., C.R.; M., L.D. A Mobile System for the Collection of Clinical Data and EEG Signals By Using The Sana Platform. *Stud. Health Technol. Inform.* **2014**, *200*, 116–123. [CrossRef]
40. Kothe, C.; Medine, D.; Boulay, C.; Grivich, M.; Stenner, T. Labstreaminglayer documentation. In *Time Synchronization*; Labstreaminglayer: San Diego, CA, USA, 2019.
41. Muller, E.; Bednar, J.A.; Diesmann, M.; Gewaltig, M.O.; Hines, M.; Davison, A.P. Python in neuroscience. *Front. Neuroinform.* **2015**, *9*, 11. [CrossRef]
42. Pernet, C.R.; Appelhoff, S.; Gorgolewski, K.J.; Flandin, G.; Phillips, C.; Delorme, A.; Oostenveld, R. EEG-BIDS, an extension to the brain imaging data structure for electroencephalography. *Sci. Data* **2019**, *6*, 103. [CrossRef]
43. Pernet, C.R.; Martinez-Cancino, R.; Truong, D.; Makeig, S.; Delorme, A. From BIDS-Formatted EEG Data to Sensor-Space Group Results: A Fully Reproducible Workflow With EEGLAB and LIMO EFG. *Front. Neurosci.* **2021**, *14*, 1407. [CrossRef]

MDPI
St. Alban-Anlage 66
4052 Basel
Switzerland
Tel. +41 61 683 77 34
Fax +41 61 302 89 18
www.mdpi.com

Applied Sciences Editorial Office
E-mail: applsci@mdpi.com
www.mdpi.com/journal/applsci

www.ingramcontent.com/pod-product-compliance
Lightning Source LLC
LaVergne TN
LVHW070658100526
838202LV00013B/991